OXFORD IB COURSE PREPARATION

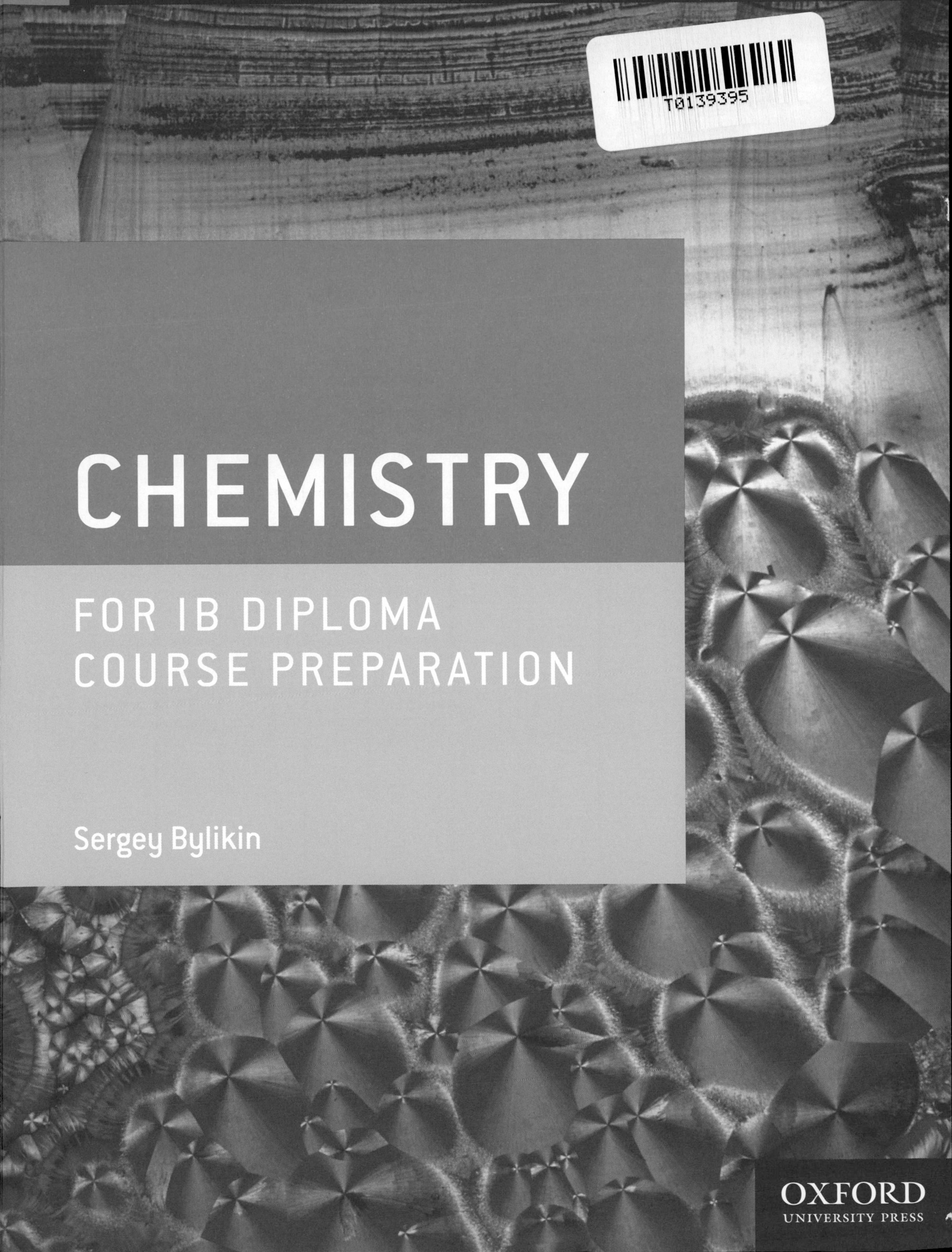



# CHEMISTRY

## FOR IB DIPLOMA COURSE PREPARATION

Sergey Bylikin

OXFORD
UNIVERSITY PRESS

# OXFORD
## UNIVERSITY PRESS

Great Clarendon Street, Oxford, OX2 6DP, United Kingdom

Oxford University Press is a department of the University of Oxford. It furthers the University's objective of excellence in research, scholarship, and education by publishing worldwide. Oxford is a registered trade mark of Oxford University Press in the UK and in certain other countries

British Library Cataloguing in Publication Data
Data available

978-0-19-842355-3

10 9 8 7 6 5 4 3 2 1

Paper used in the production of this book is a natural, recyclable product made from wood grown in sustainable forests. The manufacturing process conforms to the environmental regulations of the country of origin.

Printed in Great Britain by Bell and Bain Ltd, Glasgow

## Acknowledgements

The author would like to thank Dr Natalia Kalashnikova for her support, comments and suggestions on this book. The author would also like to thank the editing team, Helen Payne and Ben Rout, for their work on this book.

The authors and publisher are grateful to those who have given permission to reproduce the following extracts and adaptations of copyright material:

Albert Einstein: Autobiographical Notes, Open Court Publishing Centennial edition January 1999. Reproduced by permission of Cricket Media and Library of Living Philosophers, Southern Illinois University.

We have made every effort to trace and contact all copyright holders before publication, but if notified of any errors or omissions, the publisher will be happy to rectify these at the earliest opportunity.

The publisher and the authors would like to thank the following for permission to use their photographs:

**Cover:** Dr Keith Wheeler/Science Photo Library

**Artworks:** Thomson Digital

**Photos: p1:** Jan Halaska/Alamy Stock Photo; **p2 (TL):** Elena moiseeva/Shutterstock; **p2 (TB):** KPG_Payless/Shutterstock; **p2 (BR):** Georgios Kollidas/Shutterstock; **p3:** Library of Congress Prints and Photographs Division; **p5 (T):** Library of Congress Prints and Photographs Division; **p5 (B):** Library of Congress Prints and Photographs Division; **p8:** Zerbor/Shutterstock; **p10 (BL):** Mark Fergus/CSIRO; **p10 (BM):** Ben Mills/Wikimedia; **p10 (BR):** Monkey Business Images/Shutterstock; **p19:** Science Photo Library; **p26 (T):** Sputnik/Science Photo Library; **p26 (B):** Georgios Kollidas/Shutterstock; **p31:** Ann Ronan Pictures/Print Collector/ Hulton Archive/Getty Images; **p34:** Michael Ansell Photography/Shutterstock; **p39:** Science Photo Library; **p41:** Science History Images/ Alamy Stock Photo; **p58:** Keystone-France/Gamma-Keystone/Getty Images; **p70:** Valentyn Volkov/ Shutterstock; **p84:** ER Degginger/Science Photo Library; **p86:** Andrew Lambert Photography/Science Photo Library; **p89:** Sputnik/Science Photo Library; **p103:** Archive PL / Alamy Stock Photo; **p106:** Charles D. Winters/Science Photo Library; **p113:** Paul Fearn / Alamy Stock Photo; **p124:** Nagelestock.com / Alamy Stock Photo; **p138:** UtCon Collection / Alamy Stock Photo; **p142:** Ullstein Bild/Getty Images; **p149 (T):** Science History Images / Alamy Stock Photo; **p149 (B):** Art Collection 3 / Alamy Stock Photo; **p151 (TR):** Charles D. Winters/Science Photo Library; **p151 (MR):** Science Photo Library; **p151 (BR):** Wikimedia; **p161:** Charles D. Winters/Science Photo Library; **p162:** David R. Frazier Photolibrary, Inc. / Alamy Stock Photo; **p166:** Photong/ Shutterstock

# Contents

 Answers to questions in this book can be found at **www.oxfordsecondary.com/9780198423553**

# Introduction to the Diploma Programme

The **Diploma Programme** (DP) is a two-year pre-university course for students in the 16–19 age group. In addition to offering a broad-based education and in-depth understanding of selected subjects, the course has a strong emphasis on developing intercultural competence, open-mindedness, communication skills and the ability to respect diverse points of view.

You may be reading this book during the first few months of the Diploma Programme or working through the book as a preparation for the course. You could be reading it to help to decide whether the Chemistry course is for you. Whatever your reasons, the book acts as a bridge from your earlier studies to DP Chemistry, to support your learning as you take on the challenge of the last stage of your school education.

Chapters 1 through to 7 of this book explain the chemistry that you need to understand at the beginning of a DP Chemistry course. You may already have met some of this chemistry, but the book encourages you to begin to look at the concepts that underpin chemistry.

Chapter 8 of this book has advice on effective study habits and preparing for tests and examinations. Early preparation is vital even though now they may seem a long way off. The appendix lists information you will need throughout the DP Chemistry course, such as physical constants and useful equations, and a solubility table. The periodic table is printed on the back page of this book.

## DP course structure

The DP covers six academic areas, including languages and literature, humanities and social sciences, mathematics, natural sciences and creative arts. Within each area, you can choose one or two disciplines that are of particular interest to you and that you intend to study further at the university level. Typically, three subjects are studied at higher level (HL, 240 teaching hours per subject) and the other three at standard level (SL, 150 hours).

In addition to the selected subjects, all DP students must complete three core elements of the course: theory of knowledge, extended essay, and creativity, action, service.

**Theory of knowledge** (approximately 100 teaching hours) is focused on critical thinking and introduces you to the nature, structure and limitations of knowledge. An important goal of theory of knowledge is to establish links between different areas of shared and personal knowledge and make you more aware of how your own perspectives might differ from those of others.

The **extended essay** is a structured and formally presented piece of writing of up to 4,000 words based on independent research in one of the approved DP disciplines. It is also possible to write an interdisciplinary extended essay that covers two DP subjects. One purpose of the extended essay activity is to develop the high-level research and writing skills expected at university.

**Creativity, action, service** involves a broad range of activities (typically 3–4 hours per week) that help you discover your own identity, adopt the ethical principles of the IB and become a responsible member of your community. These goals are achieved through participation in arts and creative thinking (creativity), physical exercises (activity) and voluntary work (service).

## DP Chemistry syllabus

The DP Chemistry course itself is divided into four sections: Core syllabus, Additional Higher Level (AHL) material, and one of four possible Options (Materials, Biochemistry, Energy and Medicinal Chemistry) together with an **internal assessment** (IA). There is also a **group 4 project** in which all science students in a school participate.

The Chemistry course is designed so that a student can study the entire course at standard level with no prior knowledge of chemistry. At higher level, however, some earlier study of the subject is advisable.

- Chemistry standard level = Core + one Option at SL + IA + group 4 project
- Chemistry higher level = Core + AHL + one Option at HL + IA + group 4 project

Core topics cover various aspects of high school and college chemistry, including stoichiometry, atomic structure, periodicity, thermochemistry, kinetics, equilibrium and elementary organic chemistry. A separate topic introduces you to measurement, data processing, graphical techniques and spectroscopic identification of organic compounds. AHL topics expand on the content and scope of the core material. The Options contain extension material of a more in-depth and applied nature ranging from modern analytical techniques to the environmental impact of the chemical industry.

Theoretical studies are complemented by a variety of laboratory experiments that allow you to gain hands-on experience in practical chemistry and appreciate the importance of digital technology, collaboration and integrity in modern science. The collaborative nature of science is further emphasized through the interdisciplinary group 4 project, in which students of different subjects come together to analyse a common scientific or technological problem.

### Internal assessment (IA)

Chemistry is an experimental science. The ability to plan and execute an experimental project is part of your assessment, which is where the **internal assessment** (IA) comes in. The internal assessment may include oral presentations, theoretical investigations and laboratory work. About 10 hours will be devoted to the IA, probably towards the end of the course. Your teacher will support you in carrying out the IA and you will be taught the required skills throughout the course. The IA accounts for 20% of your overall examination marks. There is more advice on the IA in *8 Tips and advice on successful learning* in this book.

### Group 4 project

Most students who study a group 4 subject undertake a collaborative project within – or possibly beyond – their school. Group 4 students work together, over about 10 hours. The project is not assessed formally. It emphasizes the relationships between sciences and how scientific knowledge affects other areas of knowledge. It can be experimental or theoretical. Be imaginative in your project, and perhaps combine with a different IB World School on another continent to study a project of mutual interest.

There are ten aims addressed by every group 4 subject. Each student should:

- be challenged and stimulated to appreciate science within a global context
- develop scientific knowledge and a set of scientific techniques

- apply and use the knowledge and techniques
- develop experimental and investigative scientific skills
- learn to create, analyse and evaluate scientific information
- learn to communicate effectively using modern communication skills
- realize the value of effective collaboration and communication in science
- have an awareness of the ethical implications of science
- appreciate the possibilities and limitations of science
- understand the relationships between scientific disciplines and between science and other areas of knowledge.

## Key features of the DP Chemistry course

Chemistry is one of the core science subjects of the IB Diploma Programme. During the two years of your studies, you are encouraged to develop a broad and systematic knowledge of chemistry backed up with problem-solving skills and hands-on experience in experimental work.

The **nature of science** () is the overarching theme in all IB science subjects, including chemistry. Throughout the course, you will encounter many examples, activities and questions that go beyond the studied subject and demonstrate key principles of the scientific approach to exploring the natural world. For example, the discovery of the periodic law (*2.2 The periodic law*) was based on experimental evidence that had been accumulated by many generations of chemists and shared within the scientific community through collaboration and communication. At the same time, the success of the periodic law was a result of its predictive power and convenience for practical chemists. The discovery of atomic structure provided a theoretical explanation for periodicity and transformed the periodic law from a scientific hypothesis to a universally accepted law of nature.

Nature of science studies are not limited to the scientific method but cover many other aspects of science, from the uncertainty and limitations of scientific knowledge to the ethical and social implications of scientific research. Raising these issues will help you understand how science and scientists work in the 21st century.

**Theory of knowledge** (🧠) is another common feature of the DP Chemistry syllabus. In addition to the stand-alone theory of knowledge course taken by all DP students, much of the material in Chemistry topics can prompt wider discussions about the different ways of knowing used by scientists for interpreting experimental results. In particular, the dynamic nature of chemical equilibrium (*5.3 Chemical equilibrium*) cannot be perceived by our senses directly but can be comprehended through indirect evidence using reason and imagination.

Although theory of knowledge is not formally assessed in the DP Chemistry course, it facilitates the study of science, just as the study of science supports you in your theory of knowledge course.

**International mindedness** is one of the social aspects of science reflected in the IB mission statement, which emphasizes the importance of intercultural understanding and respect for creating a better and more peaceful world. International mindedness is actively promoted through all DP subjects by encouraging you to embrace diversity and adopt a global outlook. For example, the accelerated depletion of stratospheric ozone by chlorinated chemicals (*2.3 Chemical bonding* and *5.2 Chemical kinetics*) affected all countries around the world and has been reversed only by the combined efforts of the international community.

Various aspects of the nature of science, theory of knowledge and international mindedness can often be illustrated by the same phenomenon or event. The development of the periodic law required experimental evidence (nature of science), different ways of knowing (theory of knowledge) and international collaboration (international mindedness). Another example is the career of Fritz Haber (*5.3 Chemical equilibrium*), whose intuition and imagination (theory of knowledge) led to many discoveries that ultimately provided food for the growing world population (international mindedness) but at the same time paved the way for the industry of chemical warfare (nature of science).

Chemistry is an experimental science that provides you with numerous opportunities to develop a broad range of practical and theoretical skills. **Practical skills** (🔧) are required for setting up experiments and collecting data. Typical laboratory works are described throughout this book and include acid–base titrations (*4.2 Concentration expressions and stoichiometry*), calorimetry (*5.1 Thermochemistry*), kinetic studies (*5.2 Chemical kinetics*) and electrochemical reactions (*3.3 Redox processes*).

**Maths skills** (🖩) are needed for processing experimental data and solving problems. In addition to elementary mathematics, the IB Chemistry syllabus requires the use of powers, logarithms and p-numbers (*6.3 The concept of pH*). Working with mathematical expressions and logarithms is outlined in the appendix of this book.

**Approaches to learning** (🎓) are a variety of skills, strategies and attitudes that you will be encouraged to develop throughout the course. The Diploma Programme recognizes five categories of such skills: communication, social, self-management, research and thinking. These skills are discussed in more detail in *8 Tips and advice on successful learning* of this book.

## Assessment overview

No-one wants to think about examinations, but a clear understanding of where you are headed is important even at the start of a course.

In addition to the internal assessment discussed earlier, the **external assessment** is carried out at the end of the DP Chemistry course. Both HL and SL students are expected to take three papers as part of their external assessment. You will usually take papers 1 and 2 in one sitting with paper 3 a day or two later. The question papers are outlined in the following table:

| Paper | SL | SL timing | HL | HL timing |
|---|---|---|---|---|
| 1 | 30 multiple-choice questions on Core material | 45 minutes; 30 marks; 20% of marks | 40 multiple-choice questions on Core and AHL material | 60 minutes; 40 marks; 20% of marks |
| 2 | Short and extended written answer questions on Core material | 75 minutes; 50 marks; 40% of marks | Short written answer and extended written answer questions on Core and AHL material | 135 minutes; 95 marks; 36% of marks |
| 3 | Section A: Data-based questions and questions based on experimental work<br><br>Section B: Questions on your chosen Option | 60 minutes; 35 marks; 20% of marks | Section A: Data-based questions and questions based on experimental work<br><br>Section B: Questions on your chosen Option | 75 minutes; 45 marks; 24% of marks |

The internal and external assessment marks are combined to give your overall DP Chemistry grade, from 1 (lowest) to 7 (highest). The final score is calculated by combining grades for each of your six subjects. Theory of knowledge and extended essay components can collectively contribute up to three extra points to the overall Diploma score. Creativity, action, service activities do not bring any points but must be authenticated for the award of the IB Diploma.

## Using this book effectively

Throughout this book you will encounter separate text boxes to alert you to ideas and concepts. Here is an overview of these features and their icons:

| Icon | Feature | Description of feature |
|---|---|---|
| WE | Worked example | A step-by-step explanation of how to approach and solve a chemistry problem. |
| Q | Question | A chemistry problem to solve independently. Answers to these questions can be found at **www.oxfordsecondary.com/9780198423553** |
| 🔑 | Key term | Defines an important scientific concept used in chemistry. It is important to be familiar with these terms to prepare you for the DP Chemistry course. |
| 🧠 🧬 🎓 | DP ready | Introduces an aspect of chemistry that relates to one of three IB concepts: theory of knowledge, nature of science or approaches to learning. These concepts are explained in the *Introduction to the Diploma Programme* section of this book. |
| ∞ | Internal link | Provides a reference to somewhere within this book with more information on a topic discussed in the text, given by the section number and the topic name. For example, *2.2 The periodic law* refers to the second section in Chapter 2 of this book and covers the periodic law. |
| 🔗 | DP link | Provides a reference to a section of the DP Chemistry syllabus for further reading on a certain topic. |
| 🖩 | Maths skills | Explains an important mathematical skill required for the DP Chemistry course. |
| 🔧 | Practical skills | Relates the scientific theory to the practical aspects of chemistry you will encounter on the DP Chemistry course. |

## Linking this book to the DP Chemistry syllabus

This textbook can be read linearly, but you might find it most useful to dip into specific sections to support different areas of your learning. For example, if you are at the start of your course, you might spend some time reading *8 Tips and advice on successful learning* to develop your study skills. Alternatively, if you are learning about the mole in class, read through the parts of chapter 1 that explain the conceptual basis of the mole and practise calculations involving moles using the questions and worked examples.

The following grid shows some of the topics within the DP Chemistry syllabus, and the corresponding chapters within this book that will prepare you for the study of these topics. You can use this grid to determine which chapters you should focus your study on. For example, you can brush up on oxidation and reduction reactions by reading Chapter 3 of this book, to prepare you for *Topic 9: Redox processes* in the DP Chemistry course.

| DP Topic | Title | Sub-topics | Chapter in this book |
|---|---|---|---|
| 1 | Stoichiometric relationships | 1.1 Introduction to the particulate nature of matter and chemical change<br>1.2 The mole concept<br>1.3 Reacting masses and volumes | 1, 3, 4 |
| 2 | Atomic structure | 2.1 The nuclear atom<br>2.2 Electron configuration | 2 |
| 3 | Periodicity | 3.1 Periodic table<br>3.2 Periodic trends | 2 |
| 4 | Chemical bonding and structure | 4.1 Ionic bonding and structure<br>4.2 Covalent bonding<br>4.3 Covalent structures<br>4.4 Intermolecular forces<br>4.5 Metallic bonding | 2 |
| 5 | Energetics/ thermochemistry | 5.1 Measuring energy changes<br>5.2 Hess's Law<br>5.3 Bond enthalpies | 5 |
| 6 | Chemical kinetics | 6.1 Collision theory and rates of reaction | 5 |
| 7 | Equilibrium | 7.1 Equilibrium | 5 |
| 8 | Acids and bases | 8.1 Theories of acids and bases<br>8.2 Properties of acids and bases<br>8.3 The pH scale<br>8.4 Strong and weak acids and bases<br>8.5 Acid deposition (not covered in this book) | 6 |
| 9 | Redox processes | 9.1 Oxidation and reduction<br>9.2 Electrochemical cells | 3 |
| 10 | Organic chemistry | 10.1 Fundamentals of organic chemistry<br>10.2 Functional group chemistry | 7 |
| 14 (AHL) | Chemical bonding and structure | 14.1 Covalent bonding and electron domain and molecular geometries | 2 (very briefly) |
| 15 (AHL) | Energetics/thermochemistry | 15.2 Entropy and spontaneity | 5 (very briefly) |
| 17 (AHL) | Equilibrium | 17.1 The equilibrium law | 5 (very briefly) |
| 20 (AHL) | Organic chemistry | 20.1 Types of organic reactions | 7 (very briefly) |

# Atomic theory and stoichiometry

> *Francie came away from her first chemistry lecture in a glow. In one hour she had found out that everything was made up of atoms which were in continual motion. She grasped the idea that nothing was ever lost or destroyed. Even if something was burned up or left to rot away, it did not disappear from the face of the earth; it changed into something else — gases, liquids, and powders.*
>
> **Betty Smith,** *A Tree Grows in Brooklyn* (1943)

## Chapter context

The excerpt above, from *A Tree Grows in Brooklyn*, outlines the two most fundamental concepts of chemistry: **atomic theory** and **conservation laws**. The immense diversity of matter in our Universe originates from just over a hundred basic building blocks, the **elements**, which combine with one another in various proportions.

## Learning objectives

In this chapter you will learn about:
- → **physical** and **chemical changes**
- → the **atomic theory**
- → the nature of **chemical elements**
- → chemical **symbols**, **compounds** and **equations**
- → the **mole** and **stoichiometric relationships**.

## 🔑 Key terms introduced

- → Atoms, isotopes and chemical elements
- → The elementary charge
- → Ions, cations and anions
- → Atomic number and mass number
- → Ionic and covalent bonding
- → Chemical equations, reactants and products
- → Stoichiometry
- → Valence
- → Structural, molecular and empirical formulae
- → The mole, molar mass and mole ratio
- → Avogadro's Law
- → Standard temperature and pressure (STP)
- → The limiting reactant
- → Reaction yield
- → Green chemistry

## 1.1 The particulate nature of matter

Chemistry is the study of matter and its composition. *Matter* could be defined as anything that has mass and occupies space, while *energy* is anything that exists but does not have those properties. Matter and energy are closely associated with each other, and energy is often considered as a property of matter, such as the ability to perform work or produce heat. Although mass and energy can be converted into one another (for example, in nuclear reactors or inside the stars), chemistry studies only those transformations of matter where both mass and energy are *conserved*, that is, stay unchanged.

Matter exists in various forms. Every individual kind of matter having definite physical properties under given conditions (for example, iron, water or glucose) is called a *substance*. Substances can be isolated in pure forms, mixed together, or transformed into one another. Such transformations are called *chemical changes*, or *chemical reactions* (in contrast to *physical changes* that do not affect the identity of substances). For example, if we ignite a mixture of the two substances hydrogen and oxygen, an explosion occurs, and a new substance, water vapour, is produced. Therefore, the reaction between hydrogen and oxygen is a chemical process. In contrast, the condensation of water vapour into liquid water is a physical process, because both the vapour and the liquid are different states of the same substance, water.

**Figure 1.** Steam, liquid water and ice are the three states of water

**Figure 2.** Burning of charcoal (top) and sublimation of dry ice (bottom)

## Question

1 Charcoal, the most common form of carbon, has been used for millennia as a source of heat. When burned in air, charcoal produces carbon dioxide, which is a colourless gas under normal conditions but can also exist as a white solid (known as "dry ice") at very low temperatures. Above −78°C, dry ice sublimes without melting into gaseous carbon dioxide. Green plants use carbon dioxide and water to produce glucose and oxygen gas.

State the nature of the change (physical or chemical) for each process described in the paragraph above.

## Practical skills: Observable chemical changes

It is not always possible to distinguish between physical and chemical changes by observation. However, any spontaneous change in temperature, colour or state of matter requires a source of energy, and this energy is often chemical in nature. Observable chemical changes include spontaneous release or absorption of heat, change in colour or odour, formation of a precipitate, or release of gas (often appearing as bubbles in a liquid). Many violent processes, such as fires and explosions, are also caused by chemical changes. Finally, some chemical changes are difficult to reverse (for example, it is impossible to transform burned wood into its original state) while most physical changes are reversible.

## Key term

**Atoms** are the building blocks of all matter on Earth. There are many kinds of atom that differ in mass, size and internal structure, and they can combine in different ways.

A chemical **element** is a substance made of atoms of the same kind.

## Atoms

The building blocks of hydrogen, oxygen, water and all other chemical substances are called *atoms*. There are many kinds of atom that differ in mass, size and internal structure. A substance made of atoms of the same kind is called a *chemical element*. Currently 118 different elements are known, and this number is likely to grow in the future.

### DP ready   Theory of knowledge

#### Scientific vocabulary

The word "atom" is derived from the Greek *atomos*, which means "indivisible" and refers to a hypothesis that chemical elements cannot be split into simpler, more fundamental species. Democritus, an ancient Greek philosopher, thought

**Figure 3.** Democritus (circa 460–370 BC) on a Greek bank note

that all matter consisted of invisible, indestructible and constantly moving particles of various shapes. Although now we know that subatomic particles do exist, the term "atom" is still used by scientists when they refer to the smallest unit of a chemical element. Scientific vocabulary develops along with our understanding of the natural world, and some terms change their meaning over time.

The existence of atoms was not universally accepted until the early 1800s, when the English scientist John Dalton used his atomic theory for explaining various chemical phenomena. In particular, he suggested that the observed fact that elements combine together only in certain proportions was the result of their atomic composition. Indeed, one, two or more atoms of one element can combine with one, two or more atoms of another element, but no combination can involve a fractional number of atoms.

Based on these and other observations, Dalton's theory can be summarized as follows:

- chemical elements consist of very small particles called atoms
- all atoms of a specific element are identical but differ from atoms of other elements
- chemical compounds are formed by atoms of two or more elements in whole-number ratios
- chemical reactions involve combination, separation or rearrangement of atoms
- atoms cannot be created, subdivided or destroyed.

**Figure 4.** John Dalton (1766–1844)

**Scientific theories**

Although the ideas of Democritus and other ancient philosophers look very similar to the modern concept of atoms, their theories were not scientific, as they were derived purely by reasoning and lacked any experimental evidence. In contrast, Dalton's theory was scientific, as it provided a simple and elegant explanation for existing observations (such as the law of conservation of mass and the law of definite proportions) and had predictive power. Nevertheless, some postulates of Dalton's theory were disproved by later discoveries. This is the eventual fate of all scientific theories, which are constantly tested against experiments, and modified or replaced by new, more accurate concepts and models.

Using his theory, Dalton identified six elementary substances (hydrogen, carbon, nitrogen, oxygen, sulfur and phosphorus) and determined their *relative atomic masses* (this term is defined more precisely later in this section). To do so, he assigned the lightest atom, hydrogen, a mass of one unit and then deduced the masses of other atoms from the percentage composition of their compounds with hydrogen. Although these atomic masses were often inaccurate, Dalton was the first scientist who quantitatively described atoms in terms of their mass. To recognize the importance of this achievement, the International Union of Pure and Applied Chemistry (IUPAC) has adopted the name "dalton" for the unified atomic mass unit, which will be discussed later in this topic.

## Worked example: Deducing relative atomic masses using mass percentages

**1.** Atoms of hydrogen and oxygen combine together in a 2:1 ratio to form water. The mass percentage of hydrogen in water is 11%. Assuming that the relative atomic mass of hydrogen is 1, deduce the relative atomic mass of oxygen.

*Solution*

Let's represent the formation of water in the same way as John Dalton did over two centuries ago:

hydrogen    hydrogen    oxygen              water

If the smallest unit (molecule) of water contains two hydrogen atoms, their combined relative mass will be $2 \times 1 = 2$. This mass makes up 11% of the total mass of water.

Therefore, the relative mass of one water molecule will be $(2/11\%) \times 100\% \approx 18$.

The relative atomic mass of oxygen is therefore $18 - 2 = 16$.

Note that the ratio alone does not tell us anything about the actual number of atoms in a molecule. For example, the water molecule could contain four hydrogen and two oxygen atoms, or six hydrogen and three oxygen atoms, and so on. However, in all cases the relative atomic mass of oxygen would be the same. You can confirm this by repeating the above calculations for any larger molecule containing hydrogen and oxygen in a 2:1 ratio.

## Question

Q

2   Ammonia contains atoms of nitrogen and hydrogen in a 1:3 ratio. The mass percentage of nitrogen in ammonia is 82%. Deduce the relative atomic mass of nitrogen. (If you want to use John Dalton's notation, draw the nitrogen atom as a circle with a vertical line inside.)

Soon after Dalton formulated his theory, the relative atomic masses of many elements were determined. However, the nature of the atom remained a mystery for another century, until the scientific community accumulated enough evidence to suggest that atoms consisted of even smaller structural units.

### DP ready    Nature of science

**The scientific method**

Since the 17th century, the development of natural sciences has been based on systematic observation, measurement and experimentation. The experimental evidence was used to formulate, test and modify hypotheses, which were aimed to explain natural phenomena and predict the outcome of future experiments. The most successful hypotheses were eventually developed into scientific theories. This approach is known as the *scientific method*.

Although scientists use various approaches to obtaining, analysing and interpreting experimental evidence, they have adopted common terminology and follow a certain way of reasoning that involves deductive and inductive logic through analogies and generalizations. Mathematics, the main language of science, requires no translation and can be understood by any scientist around the world.

## Inside the atom

In 1897, the British physicist J. J. Thomson demonstrated that high-energy electric beams (so called *cathode rays*) contained negatively charged particles approximately 1,800 times lighter than a hydrogen atom. These particles, now known as *electrons* (symbol e⁻), could be emitted by any element but always had exactly the same mass and charge. Therefore, Thomson suggested that electrons were constituent parts of every atom.

Twelve years later, the New Zealand-born British physicist Ernest Rutherford and his team bombarded a piece of gold foil with positively charged α-particles, the exact nature of which was unknown at that time. While most of the particles passed through the foil without changing course, some of them were repelled or deflected by large angles. These results suggested that almost all the mass of atoms was concentrated in a small, positively charged *nucleus* surrounded by electrons. When an α-particle passed close to a nucleus, it was repelled or deflected. However, the small atomic nuclei were easy to miss, so the majority of α-particles passed through the foil without any resistance.

In 1911, Rutherford summarized the results of his experiments by proposing the *planetary model of the atom*, also known as the *Rutherford model* (figure 7). In this model, negatively charged electrons orbit the positively charged atomic nucleus in the same way as planets orbit the Sun. Just as the Sun contains 99.8% of the solar system's mass, the atomic nucleus contains over 99.9% of the mass of the entire atom. However, instead of gravity, electrostatic attraction holds the electrons around the nucleus.

**Figure 5.** J. J. Thomson (1856–1940)

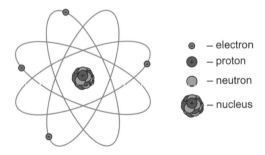

- electron
- proton
- neutron
- nucleus

**Figure 7.** The Rutherford model of the atom

**Figure 6.** Ernest (Lord) Rutherford (1871–1937)

Rutherford was able to estimate the diameter of the atomic nucleus. He found that it was about $10^{-14}$ m, or 10,000 times smaller than the atomic diameter (about $10^{-10}$ m). If we could enlarge the atom to the size of a football stadium, the nucleus would look like a golf ball in the middle of the field.

**DP ready    Nature of science**

    **Internal link**

### Wave–particle duality

According to modern theories, the electrons in an atom do not orbit the nucleus but rather fill the whole volume of the atom as diffuse clouds. By doing so, electrons demonstrate the properties of particles and waves at the same time. This peculiar behaviour could not be explained by the planetary theory, which was eventually replaced by the quantum model of the atom. Nevertheless, the main concept of the Rutherford model – a small, dense, positively charged nucleus surrounded by negatively charged electrons of negligible mass – has been confirmed by various experiments and still remains in use as an excellent visualization of the atomic structure.

You can read about the quantum model of the atom in **2 Electron structure and chemical bonding**.

Later studies showed that atomic nuclei contain two types of subatomic particle: positively charged *protons* (symbol p) and uncharged *neutrons* (symbol n). Both particles have approximately the same mass (about $1.67 \times 10^{-27}$ kg) while the mass of the electron is over 1800 times lower, about $9.11 \times 10^{-31}$ kg. These masses are extremely small: a proton is as many times lighter than a golf ball as the golf ball is lighter than the Earth.

Since the masses of atoms and subatomic particles are so small, they are awkward to work with and take a lot of space when written. Therefore, they are often expressed in *unified atomic mass units* (symbol u), or *daltons* (symbol Da):

$1 \text{ u} = 1 \text{ Da} \approx 1.66 \times 10^{-27}$ kg

$m(\text{p}) \approx m(\text{n}) \approx 1.01$ u

$m(\text{e}^-) \approx 0.000549$ u

The unified atomic mass unit is traditionally used in chemistry, while the dalton is more common in biochemistry and molecular biology, especially for expressing the masses of large organic molecules. In this book, we will be using the unified atomic mass unit.

**DP link**

You will learn why the mass of a proton is not exactly 1 u if you study **C.7 Nuclear fusion and nuclear fission** in the IB Chemistry Diploma Programme.

In simple calculations, the masses of protons and neutrons are often rounded to 1 u each, and the masses of electrons in atoms are considered negligible. For example, an atom of sodium contains 11 protons, 12 neutrons and 11 electrons, so its mass can be found as $11 \times 1 + 12 \times 1 + 11 \times 0 = 23$ u. The exact mass of this atom determined experimentally is 22.99 u, which is very close to our approximate value.

### Question

3  An atom of aluminium contains 13 p, 14 n and 13 e⁻. Calculate the mass of this atom in u and kg.

**Key term**

The electric charge carried by a single electron is known as the **elementary charge** (*e*) and has a value of approximately $1.602 \times 10^{-19}$ C. The charges of subatomic particles are commonly expressed in elementary charge units. For example, the charge of an electron can be represented as −*e*, and the charge of a proton as +*e*. The symbol *e* is often omitted, so it is customary to say that an electron and proton have charges of −1 and +1, respectively.

*Relative atomic masses* ($A_r$) are numerically equal to the masses of atoms in u but have no units. For example, the $A_r$ of the sodium atom is 22.99. Note that John Dalton assumed that hydrogen had $A_r = 1$ while in modern chemistry it has an $A_r$ of 1.01. The reason for this will become clear in *1.3 Stoichiometric relationships*, when we will introduce the concept of the mole.

The electric charges of the proton and the electron have opposite signs but identical magnitude, the *elementary charge*. Since the atom as a whole is electrically neutral, it must contain the same number of protons and electrons. If an atom loses one or more electrons, it develops a net positive charge and becomes a *cation*. Similarly, if an atom gains one or more electrons, it develops a net negative charge and becomes an *anion*. Cations and anions, collectively known as *ions*, are often formed in chemical reactions, so the number of electrons in a given atomic species can vary.

However, chemical changes do not affect atomic nuclei, so the numbers of protons and neutrons in a given atomic species remain constant. The number of protons in the nucleus, also known as the *atomic number* (*Z*), is numerically equal to the nuclear charge, as each proton has a charge of +1. In turn, the nuclear charge defines the chemical properties of the atom, such as its ability to lose, gain or share electrons and interact with other species.

The atomic number is the most fundamental characteristic that distinguishes one chemical element from another. All atoms of a given element have the same number of protons in their nuclei and thus the same atomic number. In contrast, the number of neutrons ($N$) can vary among the nuclei of the same element, as neutrons do not affect the nuclear charge and thus have very little effect on the chemical properties of the atom.

### The elements

In chemistry, the elements are represented by their *symbols*, which consist of one or two letters and are derived from the element names. For example, the chemical symbol for hydrogen is H (the first letter of *h*ydrogen), and the chemical symbol for iron is Fe (the first two letters of the Latin *ferrum*, iron). Chemical elements and their symbols are listed in table 1 and in the periodic table at the back of this book.

If you look closely at the fourth column of table 1, you will notice that most $A_r$ values are very close to whole numbers. For example, $A_r(H) = 1.01 \approx 1$ and $A_r(Na) = 22.99 \approx 23$. Some elements (Mg, Cl, Fe and Cu), however, do not obey this rule: their $A_r$ values differ significantly from whole numbers. Such elements consist of several *isotopes*. For example, chlorine has two common isotopes with relative atomic masses of 35 and 37. In naturally occurring chlorine, these two isotopes are mixed in a ratio that produces the *average* relative atomic mass of 35.45.

**Key term**

**Ions** are atoms that have lost or gained one or more electrons and thus have a net charge. **Cations** have a positive charge (have lost electrons), and **anions** have a negative charge (have gained electrons).

The number of protons in the nucleus is the **atomic number** ($Z$).

**Key term**

**Isotopes** are atoms with the same number of protons but different numbers of neutrons in their nuclei.

The **mass number** ($A$) is the total number of protons and neutrons in the nucleus.

**Table 1.** Common chemical elements

| Symbol | Name | Atomic number ($Z$) | Relative atomic mass ($A_r$) | Notes |
|---|---|---|---|---|
| H | hydrogen | 1 | 1.01 | |
| He | helium | 2 | 4.00 | |
| C | carbon | 6 | 12.01 | |
| N | nitrogen | 7 | 14.01 | |
| O | oxygen | 8 | 16.00 | |
| F | fluorine | 9 | 19.00 | |
| Na | sodium | 11 | 22.99 | Latin *natrium* |
| Mg | magnesium | 12 | 24.31 | |
| Al | aluminium | 13 | 26.98 | |
| P | phosphorus | 15 | 30.97 | |
| S | sulfur | 16 | 32.07 | |
| Cl | chlorine | 17 | 35.45 | |
| K | potassium | 19 | 39.10 | Latin *kalium* |
| Ca | calcium | 20 | 40.08 | |
| Fe | iron | 26 | 55.85 | Latin *ferrum* |
| Cu | copper | 29 | 63.55 | Latin *cuprum* |

The total number of protons and neutrons in the nucleus is known as the *mass number* ($A$). To distinguish between isotopes, their mass numbers are often written as superscript indices to the left of the element symbol. For example, chlorine-35 can be represented as $^{35}Cl$, and chlorine-37 as $^{37}Cl$. It is also common to show the atomic number ($Z$) as a subscript index directly beneath the mass number. Since each atom of chlorine has 17 protons in its nucleus, chlorine-35 and chlorine-37 can be represented as $^{35}_{17}Cl$ and $^{37}_{17}Cl$, respectively. We can always find the number of neutrons ($N$) in the nucleus as the difference between its mass number and atomic number:

$$N = A - Z$$

For example, $^{35}_{17}Cl$ has $35 - 17 = 18$ neutrons, and $^{37}_{17}Cl$ has $37 - 17 = 20$ neutrons.

**Figure 8.** Smoke detectors contain radioactive isotopes

**DP link**

Radioactive materials are used in industry, medicine and even household devices, such as smoke detectors. These and many other applications are discussed in **C.3 Nuclear fusion and fission** and **D.8 Nuclear medicine** in the IB Chemistry Diploma Programme.

---

**DP ready** **Nature of science**

### Radioactivity

Isotopes are known for all chemical elements. Stable isotopes can exist indefinitely, while *radioisotopes* decompose with time. This decomposition, known as *radioactivity* or *radioactive decay*, can produce a variety of high-energy particles, including electrons, neutrons, photons and nuclei of other elements. For example, the α-particles used in Rutherford's experiments were later identified as nuclei of helium-4, each consisting of two protons and two neutrons ($^{4}_{2}He$). They are produced by the radioactive decay of radium-226 as follows:

$$^{226}_{88}Ra \rightarrow {}^{4}_{2}He + {}^{222}_{86}Rn$$

The second product in the above *nuclear equation*, the gas radon-222, is also radioactive. Since radium-226 is naturally present in many minerals, radon-222 is constantly released from soil and may accumulate in poorly ventilated buildings. The levels of radon in particularly affected areas are monitored by authorities and ordinary people, who often keep portable radiation detectors in their homes.

---

### Electrons and ions

So far, our study of the atom has been focused on the nucleus. It is now time to have a closer look at the electrons, which occupy 99.9999999999% of the atom's volume and are largely responsible for any chemical changes.

As you know already, the number of electrons in an atom is equal to the number of protons in its nucleus. For example, the nucleus of a sodium atom contains 11 protons, so this atom must have 11 electrons. In chemistry, atoms are represented by the same symbols as elements, so Na may refer to either a single sodium atom or, more generally, sodium as a type of atom.

When an atom loses electrons, it becomes an ion with a net positive charge (cation). This charge must be shown as a superscript index to the right of the element symbol. For example, sodium atoms readily lose one electron to produce sodium cations, $Na^+$:

$$Na \rightarrow Na^+ + e^-$$

Similarly, magnesium atoms lose two electrons each, producing doubly charged magnesium cations:

$$Mg \rightarrow Mg^{2+} + 2e^-$$

The reason why some elements produce singly charged ions while others tend to form multiply charged ions will be discussed in the next chapter. For now, it is sufficient to understand that each element tends to lose, gain or share only a certain number of electrons. As you will see, this ability explains the fact that elements combine with one another in definite proportions.

Monoatomic cations have the same names as the elements from which they are formed. When it is clear that we are talking about a cation, the words "cation" or "ion" can be omitted, so "magnesium" can refer to an atom, a chemical element, or a magnesium cation.

Cations may also consist of more than one atom. Such *polyatomic*, or *molecular, cations* often have common names ending with the suffix "-onium" (for example, the ammonium ion, $NH_4^+$).

Some elements, such as chlorine and sulfur, tend to gain electrons from other atoms and produce negatively charged ions (anions):

$$Cl + e^- \rightarrow Cl^-$$
$$S + 2e^- \rightarrow S^{2-}$$

Monoatomic anions are named by combining a part of the element name with the suffix "-ide", as shown in table 2.

 **Internal link**

The charges on ions are examined in
**2 Electron structure and chemical bonding.**

**Table 2.** Common monoatomic anions and their names

| Element | | Anion | |
|---|---|---|---|
| **Symbol** | **Name** | **Symbol** | **Name** |
| H | hydrogen | $H^-$ | hydride |
| N | nitrogen | $N^{3-}$ | nitride |
| O | oxygen | $O^{2-}$ | oxide |
| F | fluorine | $F^-$ | fluoride |
| P | phosphorus | $P^{3-}$ | phosphide |
| S | sulfur | $S^{2-}$ | sulfide |
| Cl | chlorine | $Cl^-$ | chloride |
| Br | bromine | $Br^-$ | bromide |
| I | iodine | $I^-$ | iodide |

The names of polyatomic anions usually end with "-ate" (for example, the sulfate ion, $SO_4^{2-}$) or "-ite" (for example, the sulfite ion, $SO_3^{2-}$).

If we need to show the ionic charge of a specific isotope, all indices can be added to a single chemical symbol. For example, the anion of sulfur-32 can be represented as $^{32}_{16}S^{2-}$ or $^{32}S^{2-}$.

 **Internal link**

Polyatomic anions are discussed in more detail in
**3 Inorganic chemistry.**

**Question**

4  Complete the table below by deducing the composition, names and/or symbols of atoms and ions.

| Symbol | Name | Mass number ($A$) | Number of | | |
|---|---|---|---|---|---|
| | | | neutrons | protons | electrons |
| | hydrogen cation | | 0 | | |
| | | 1 | | | 1 |
| | hydride | | 0 | | |
| $^{27}_{13}Al^{3+}$ | | | | | |
| | chloride | 35 | | | |
| $^{14}_{7}N^{3-}$ | | | | | |

Before moving on to the next topic, check that you have solved all the problems in the text and developed a working knowledge of atomic structure. You are also advised to review the information given in tables 1 and 2. Although you do not have to memorize the symbols and names of all elements, by now you should be able to recognize at least some of them, such as hydrogen (H), oxygen (O), sodium (Na) and chlorine (Cl), without checking the book.

## 1.2 Chemical substances, formulae and equations

Atoms and ions are the smallest units of matter that still possess certain chemical properties. While these species can exist individually, they tend to combine together and form chemical substances.
*Elementary substances* contain atoms of a single element while *chemical compounds* contain atoms of two or more elements bound together by chemical forces. For example, magnesium metal is an elementary substance, as it contains only one type of atom, Mg. Similarly, sulfur (S) is another elementary substance composed of sulfur atoms only. In contrast, magnesium sulfide is a chemical compound, as it consists of two different, chemically bound atomic species, Mg and S (figure 9).

**Figure 9.** Magnesium (left), sulfur (middle) and magnesium sulfide (right)

### Practical skills: Mixtures and pure substances

If we grind magnesium metal and sulfur into fine powders and mix them together at room temperature, both elements will retain their chemical identities. For example, we will still see shiny particles of magnesium and yellow crystals of sulfur in this mixture under a microscope. Moreover, we will be able to separate this mixture into individual substances by shaking it with water: sulfur will float on the surface of water while magnesium will sink to the bottom. Finally, the mixture of magnesium and sulfur has variable composition, as we can mix these substances together in any proportion.

However, if we heat up the mixture of magnesium and sulfur, a vigorous reaction occurs, and the mixture turns into a white powder of magnesium sulfide. It will look like white crystalline material under a microscope, and we will not be able to isolate magnesium and sulfur from this compound by any physical methods. Therefore, magnesium sulfide is a pure substance. It has a definite composition (equal number of magnesium and sulfur atoms), and its properties differ from those of both the original substances (magnesium and sulfur) and their mixture.

Mixtures in which the components are clearly visible are called *heterogeneous* (from the Greek *heteros* "other"). The mixture of magnesium and sulfur is an example of a heterogeneous mixture. Other mixtures can be blended so well that they have uniform composition, and every part of the mixture has exactly the same properties. Such mixtures are called *homogeneous* (from the Greek *homos* "same"). Air, sea water and most metal alloys are examples of homogeneous mixtures.

**Internal link**

The most common homogeneous mixtures, aqueous solutions, will be discussed in **4 Solutions and concentration**.

### Types of bonding: Ionic bonding

There are several ways in which atomic species can combine together. The first and probably the most obvious way is the electrostatic attraction between oppositely charged ions, which is known as *ionic bonding*. For example, magnesium sulfide consists of magnesium cations, $Mg^{2+}$, and sulfide anions, $S^{2-}$. In solid magnesium sulfide, these ions form a regular, highly symmetrical arrangement, known as the *crystal lattice*, in which each ion is surrounded by ions of the opposite charge (figure 10). Other ionic compounds, such as sodium chloride (table salt), have similar structures.

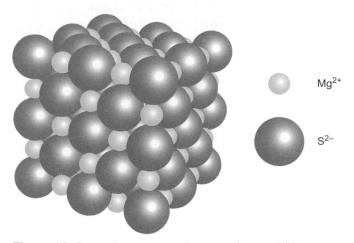

$Mg^{2+}$

$S^{2-}$

**Figure 10.** Crystal structure of magnesium sulfide

The formation of magnesium sulfide can be represented by the following *chemical equation*:

$$Mg + S \rightarrow MgS$$

**Key term**

**Ionic bonding** is the electrostatic attraction between oppositely charged ions that holds them together in a compound.

In a **chemical equation**, substances are represented by symbols for the elements in their **chemical formulae**. The **reactants** are shown on the left, followed by a **reaction arrow** that shows that a change has occurred and then the **products** of the reaction are shown on the right.

In this equation, the elementary substances (magnesium metal and sulfur) are represented by the same symbols as their chemical elements (magnesium and sulfur, respectively). Since these substances undergo chemical changes by reacting with each other, they are called *reactants*. In chemical equations, reactants are traditionally placed on the left and followed by a *reaction arrow* that shows the direction of the reaction. The arrow points at the *reaction product(s)* formed as a result of the chemical changes. In our case, the reaction product is magnesium sulfide, MgS.

The *chemical formula* MgS represents a chemical compound of a definite composition (magnesium and sulfide ions combined in 1:1 ratio). This formula also shows that magnesium sulfide is electrically neutral, as positive charges of $Mg^{2+}$ ions in this compound are compensated by exactly the same number of negative charges of $S^{2-}$ ions. If needed, the ionic nature of magnesium sulfide can be shown explicitly as $Mg^{2+}S^{2-}$. However, in chemical formulae of neutral species, the ionic charges of individual elements are usually omitted.

It is important to understand that any chemical compound or elementary substance must be electrically neutral as a whole. The ions of the same sign repel one another, so they cannot stay together in any significant quantity unless their charges are cancelled by the ions of the opposite sign. If we could somehow place 1 g of only $Mg^{2+}$ ions together, their electrostatic repulsion would produce an explosion more powerful than the simultaneous detonation of all nuclear weapons in the world! The need to balance ionic charges is one of the reasons why chemical elements combine with one another in definite proportions.

Ions of different charge magnitudes can form compounds together. For example, sodium atoms produce singly charged ions, $Na^+$, while sulfur atoms form doubly charged ions, $S^{2-}$. To balance these charges, we need twice as many $Na^+$ ions as $S^{2-}$ ions. Therefore, sodium and sulfur combine together in a 2:1 ratio in sodium sulfide:

**Key term**

The term **stoichiometry** is derived from two Greek words, *stoicheion* "element" and *metron* "measure", and refers to the relationship between the amounts of substances participating in a chemical reaction. In a broader sense, this term is used to describe any quantitative relationships in chemistry, such as percent composition of compounds, reaction yields, volume ratios and so on; see **1.3 Stoichiometric relationships**.

$$2Na + S \rightarrow Na_2S$$

In this equation, the *stoichiometric coefficient* 2 placed before the symbol of the sodium atom shows that two atoms of sodium react with each sulfur atom. (The stoichiometric coefficient 1 before S and $Na_2S$ is omitted, as it would normally be omitted in any other expression, such as "$2x + y = z$".) Note that the expression "$2Na + S$" shows only the ratio of the atoms, not their exact quantities: it might be ten billion sodium atoms and five billion sulfur atoms.

The *subscript index* 2 placed after the symbol of sodium in the formula of sodium sulfide, $Na_2S$, has a slightly different meaning. This index shows the exact number of $Na^+$ ions in the smallest structural unit of sodium sulfide. While the stoichiometric coefficients (numbers before chemical symbols and formulae) can vary among equations, the subscript indices for each particular compound must always stay the same. This distinction will become clear as you study more examples.

To deduce a formula of an ionic compound, all we need to know is the charges of the ions involved. Most of these charges can be deduced from the periodic table, which we will discuss in the next chapter. For now, it is sufficient to know that the cations of sodium ($Na^+$) and potassium ($K^+$) are singly charged, magnesium ($Mg^{2+}$) and calcium ($Ca^{2+}$) form doubly charged cations, and aluminium forms a triply charged cation ($Al^{3+}$). The charges and names of all common anions are listed in table 2.

---

### Worked example: Deducing formulae of ionic compounds

**2.** Write the formula of calcium phosphide. (This compound will contain ions of $Ca^{2+}$ and $P^{3-}$.)

*Solution*

To produce a neutral compound, the total charge of $Ca^{2+}$ ions must be exactly countered by the total charge of $P^{3-}$ ions.

The lowest common multiple of 2 and 3 is 6, so we need three $Ca^{2+}$ ions (2 × 3 = 6 positive charges) and two $P^{3-}$ ions (3 × 2 = 6 negative charges).

Therefore, the formula of calcium phosphide is $Ca_3P_2$.

---

### Question

5 Complete the table below by deducing the formulae and names of ionic compounds. One cell is already filled as an example.

| | $Cl^-$ | $O^{2-}$ | $N^{3-}$ |
|---|---|---|---|
| $Na^+$ | | | |
| $Ca^{2+}$ | | CaO<br>calcium oxide | |
| $Al^{3+}$ | | | |

---

### Worked example: Balancing equations

**3.** Balance the equation for the production of calcium phosphide above.

*Solution*

If we want to represent the formation of calcium phosphide by a chemical equation, we must not forget about the stoichiometric balance: the number of atoms of each kind on both sides of the equation must be the same. For example, the following equation is not balanced:

$$Ca + P \rightarrow Ca_3P_2 \qquad \text{(incorrect!)}$$

It contains one Ca atom on the left but three Ca atoms on the right. Similarly, there is one P atom on the left but two P atoms on the right.

To balance an equation, we can change the stoichiometric coefficients before any substances (reactants and/or products) but not the subscript indices that show the ratio of the atoms in chemical formulae of compounds.

Start by balancing calcium by placing the coefficient 3 before Ca on the left:

$$3Ca + P \rightarrow Ca_3P_2 \qquad \text{(still incorrect!)}$$

Now calcium is balanced, but phosphorus is not. We can correct it by placing 2 before P on the left:

$$3Ca + 2P \rightarrow Ca_3P_2 \qquad \text{(correct)}$$

Both elements are now balanced, so the equation is complete.

It is good practice to balance all chemical equations you write, as it will help you to avoid errors in stoichiometric calculations (see *1.3 Stoichiometric relationships*).

**Question**

6   Write and balance chemical equations for the formation of each ionic compound from question 5.

**Key term**

**Covalent bonding** occurs when a pair of electrons is shared between two atoms. The electrostatic attraction between the nuclei of each atom and these electrons holds the atoms together.

The term **covalent** is related to the word **valence**, which in turn is derived from Latin valentia ("power"). In chemistry, *valence* is the ability of atoms to form **covalent bonds** with one another. An atom is said to be **monovalent** if it can form only one covalent bond; **divalent** atoms form two covalent bonds, and so on.

**Key term**

In **structural formulae**, each covalent bond is shown as a line between atom symbols.

**Molecular formulae** show the number of atoms in a molecule, but not necessarily how they are joined.

## Types of bonding: Covalent bonding

Although ionic compounds are very common in chemistry, there are other types of chemical bonding. *Covalent compounds* are formed when atoms share their electrons rather than transferring them completely from one element to another. For example, the lightest element, hydrogen, exists in the form of a diatomic *molecule*, $H_2$, where the two hydrogen atoms have a common pair of electrons. If we draw electrons as dots near the element symbols, the formation of $H_2$ from two hydrogen atoms can be represented by the following scheme:

$$H\cdot + \cdot H \rightarrow H\!:\!H$$

Electrons are charged negatively while the atomic nuclei have positive charges. When a pair of electrons is shared between two atoms, both nuclei are electrostatically attracted to these electrons. In other words, shared electrons act as a "glue" that holds the atoms together. This type of chemical bonding is called *covalent bonding*, and each pair of shared electrons is treated as a separate *covalent bond*.

A hydrogen atom has a single electron, so it is *monovalent* in all its compounds. Another element that is usually monovalent is chlorine. Oxygen is *divalent*, and nitrogen is usually *trivalent*. All these elements form stable diatomic molecules with one or more shared electron pairs. Covalent molecules are often represented by *structural formulae*, where each covalent bond is shown as a line (figure 11).

**Figure 11.** Structural formulae of some diatomic molecules

Since these diatomic molecules are the smallest structural units of elementary hydrogen, chlorine, oxygen and nitrogen, their formulae in chemical equations must be written as $H_2$, $Cl_2$, $O_2$ and $N_2$, respectively, rather than H, Cl, O and N. Such formulae are known as *molecular formulae*, as they show the exact number of atoms in each molecule. For example, the reaction between hydrogen and chlorine is represented by the following equation:

$$H_2 + Cl_2 \rightarrow 2HCl$$

The product of this reaction, hydrogen chloride, HCl, is another covalent molecule that contains a single covalent bond between hydrogen and chlorine. Generally, the formulae of covalent compounds are constructed in the same way as the formulae of ionic compounds, except that the valences of atoms are used instead of ionic charges.

For example, water can be produced by the reaction of hydrogen with oxygen:

$$H_2 + O_2 \rightarrow water$$

Therefore, we can expect that a molecule of water consists of H and O atoms. Since oxygen is divalent, it will form bonds with two monovalent hydrogen atoms:

H—O—H

Thus, the formula of water is $H_2O$. Now we can complete and balance the equation:

$$2H_2 + O_2 \rightarrow 2H_2O$$

If the valence of an element is unknown, we can often deduce it from the formula of a covalent compound. For example, the major component of natural gas, methane, has the formula $CH_4$. Since hydrogen is monovalent, carbon will need to form four covalent bonds to hold four hydrogen atoms. Therefore, carbon is tetravalent, and the formula of methane can be drawn as follows:

$$
\begin{array}{c}
\quad H \\
\quad | \\
H - C - H \\
\quad | \\
\quad H
\end{array}
$$

Methane is the simplest *organic compound*, as it contains both carbon and hydrogen atoms. We will discuss its structure and properties in the last chapter of this book.

## Question

7  Draw the structures of the following covalent molecules, representing each covalent bond as a line: HF, $H_2S$, $NH_3$, $CF_4$, $CO_2$.

8  For each compound in Question 7, write and balance a chemical equation where this compound is formed from elementary substances.

### DP ready  Nature of science

**Trivial and systematic names**

Many covalent compounds have trivial names, such as water ($H_2O$), ammonia ($NH_3$) and silica ($SiO_2$). These names are so common that they are likely to stay in use for a very long time, both in chemical literature and everyday life. Some of these trivial names are introduced throughout this book. However, scientists also have *systematic names* for these substances, which are constructed in the same way as those for ionic compounds:

$H_2O$ – hydrogen oxide

$NH_3$ – hydrogen nitride

$SiO_2$ – silicon oxide

 **Internal link**

You will learn more about systematic nomenclature of inorganic compounds in **3 Inorganic chemistry**.

## Types of bonding: Metallic bonding

In addition to ionic and covalent compounds, atoms of certain elements combine together by *metallic bonding*. As suggested by the name, this type of bonding usually occurs in metals. Atoms of metals lose their electrons easily, producing cations. The free electrons are distributed throughout the whole volume of the metal, holding the cations together by electrostatic attraction. This type of bonding is similar to both ionic bonding (as it involves ions) and covalent bonding (as the electrons act as a "glue" between the atoms). However, in contrast to ionic bonding, metals contain only positively charged ions (cations), with the negative charges of the free-moving electrons balancing the charges on the cations. And in contrast to covalent bonds, which are *localized* between specific atoms, the metal bonding is *delocalized*, so all metal cations are bonded to one another by a single cloud of shared electrons.

In chemical equations, metallic substances are represented by their element symbols without any subscript indices (for example, Na or Fe), as the total number of metal atoms in a sample can vary. If several metals are present in the same sample, their atoms may share electrons with one another and form an *alloy*. Since alloys have variable composition, they are not considered chemical compounds but rather mixtures or solid solutions.

Further discussion of ionic, covalent and metallic bonding requires more detailed knowledge of the electron structure of chemical elements, which you will explore in the next chapter. Then, in the same chapter, you will see how the type of chemical bonding in various compounds can be predicted, and how it affects the properties of these compounds.

## States of matter

So far, we have considered individual atoms and molecules without paying much attention to the physical properties of substances. It is now time to review the states of matter and the way they are represented in chemical formulae and equations.

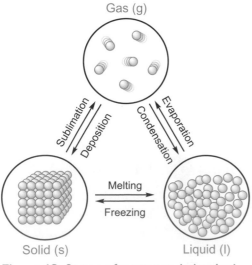

**Figure 12.** States of matter and physical changes

As you probably know, the three most common states of matter are solid, liquid and gaseous (figure 12). In solid substances, every atom, ion or molecule occupies a certain position owing to strong attractive forces between these particles. Solids are almost incompressible, and have definite volume and shape. In liquids, the forces between particles are weaker, so atoms, ions or molecules can move around but cannot get far away from each other. As a result, liquids are also almost incompressible and have definite volume but no definite shape. In gases, the attractive forces between particles are negligible, so atoms or molecules move freely and generally faster than in liquids. Gases can be compressed and have no definite volume or shape, so they fill the whole volume of the vessel.

To specify the state of a substance, we show the first letter of the state ("s" for solid, "l" for liquid and "g" for gaseous) in brackets after the formula, for example:

$$H_2O(s) \qquad H_2O(l) \qquad H_2O(g)$$

solid water      liquid water      gaseous water

(ice)      (water vapour or steam)

Using these state symbols, we can represent physical changes in the same way as chemical. For example, evaporation of liquid water can be shown as follows:

$$H_2O(l) \rightarrow H_2O(g)$$

A special symbol, "(aq)", is used for substances or ions in aqueous solutions. For example, the expression "NaCl(aq)" tells us that sodium chloride is dissolved in water while "NaCl(s)" refers to the individual compound (solid sodium chloride). Please note the difference between liquid and aqueous states: the expression "NaCl(l)" means that sodium chloride is molten and probably has a very high temperature, as this salt melts at 801°C (1074 K).

In chemical equations, states are particularly important when we discuss reactions in aqueous solutions, thermal effects, reaction rates and chemical equilibrium. Many substances behave differently depending on whether they are in their pure form or dissolved in water, so we need to know their states to write the correct chemical equation. In addition, the states of reactants and products can provide some information about the reaction conditions and observable chemical changes, such as the release of a gas or the formation of a precipitate, for example:

$$Mg(s) + 2HCl(aq) \rightarrow MgCl_2(aq) + H_2(g)$$

The above equation tells us that solid magnesium metal reacted with an aqueous solution of hydrogen chloride, producing bubbles of a gas (hydrogen) and an aqueous solution of magnesium chloride. We can also predict that the amount of magnesium decreased and, if the reaction went to completion, the metal eventually disappeared. Finally, the equation tells us that magnesium chloride is soluble in water, as otherwise it would produce a solid precipitate, $MgCl_2(s)$.

Showing states of all reactants and products is good practice, and in this book we will use state symbols in most equations; any exceptions will be explained in the text.

### Question

9 Using equations and state symbols, represent all physical and chemical changes described in question 1 on page 2. The formula of glucose is $C_6H_{12}O_6$.

**Internal link**

Reactions in aqueous solutions are discussed later, in **4 Solutions and concentration**. Thermal effects, reaction rates and chemical equilibrium are all covered in **5 Thermochemistry, kinetics and equilibrium**.

You have now learned how the composition and transformations of chemical substances can be represented by chemical formulae and equations. In the next topic, you will see how these formulae and equations can be used for quantitative description of matter and stoichiometric calculations.

## 1.3 Stoichiometric relationships

Atoms and molecules are so small that their masses cannot be measured directly. Even a million atoms of lead, Pb, the heaviest stable element, would have a mass of only $3.4 \times 10^{-16}$ g, which is too small to be weighed even on the most sensitive laboratory balance. At the same time, the number of Pb atoms in 1 g of lead is a huge number, about $2.9 \times 10^{21}$, which is hard to imagine, let alone count. Therefore, chemists need a unit allowing them to work comfortably with both very small masses and very large numbers of atoms. This unit, the *mole*, was devised in the 19th century and quickly became one of the most useful concepts in chemistry.

### The mole

In 1971, the 14th General Conference of Weights and Measures adopted the mole (symbol "mol") as one of the base SI units.

> **Key term**
>
> The **mole** was defined as the *amount of substance* that contains as many elementary entities (atoms, molecules, ions, electrons or other particles) as there are atoms in 0.012 kg (or 12 g) of carbon-12.

**DP ready**  **Nature of science**

**The International System of Units**

The International System of Units (SI, abbreviated from French *Système international d'unités*) is the system of measurement most widely used by scientists. Its building blocks are the seven base units: length (metre, m), mass (kilogram, kg), time (second, s), electric current (ampere, A), temperature (kelvin, K), amount of substance (mole, mol) and luminous intensity (candela, cd). All other units, such as those of volume ($m^3$), density ($kg\,m^{-3}$), energy (joule, J, where $1\,J = 1\,kg\,m^2\,s^{-2}$) and so on, are derived from the seven base units. The use of universal and precisely defined units is extremely important, as it allows scientists from different countries to understand one another and share the results of their discoveries.

Base and derived SI units, physical constants and useful expressions are listed in the appendix.

The number of atoms in 12 g of $^{12}_{6}C$ is known with very high precision. Its current value is $6.022140857 \times 10^{23}$, but it might be revised slightly in the future. In all our calculations, we will use the rounded value $6.02 \times 10^{23}$.

The *Avogadro constant* ($N_A$) is the conversion factor linking number of particles and number of moles. It has the unit of $mol^{-1}$:

$$N_A = 6.02 \times 10^{23}\,mol^{-1}$$

In chemical calculations, the Avogadro constant is used in the same way as any other conversion factor. For example, to convert kilograms into grams, we need to multiply the mass in kg by 1000. Similarly, to convert the amount of substance ($n$) into the number of atoms or any other structural units ($N$), we need to multiply that amount by $N_A$:

$$N = n \times N_A$$

> **Worked example: Using the Avogadro constant**  **WE**
>
> **4.** Calculate the amount of lead in a sample containing $2.9 \times 10^{21}$ atoms of this element.
>
> *Solution*
>
> $$N = n \times N_A \quad \text{rearrange to } n = \frac{N}{N_A}$$
>
> $$n = 2.9 \times 10^{21}/6.02 \times 10^{23}\,mol^{-1} \approx 0.0048\,mol = 4.8\,mmol.$$

## Question

10 Calculate:
   a) the number of atoms in 2.5 mol of sodium metal;
   b) the number of molecules in 0.25 mol of water;
   c) the number of atoms in 0.25 mol of water.

**DP ready** | **Theory of knowledge**

### Mole Day

Every year, October 23 (written as 10/23 in the USA) is celebrated by many chemists as Mole Day. The event begins at 6.02am and ends at 6.02pm Combined together, the time and date resemble the Avogadro constant, $6.02 \times 10^{23}$ mol$^{-1}$. This constant is named after the Italian scientist Amedeo Avogadro, who made a significant contribution to chemistry by showing that equal volumes of gases under the same conditions contain equal numbers of molecules.

Schools and other organizations around the world celebrate Mole Day with various activities related to the mole with the aim to foster public interest in chemistry.

**Figure 13.** Amedeo Avogadro (1776–1856)

### Calculations involving chemical formulae

By definition, the *molar mass* of carbon-12 is exactly 12 g mol$^{-1}$ (as 12 g of this isotope contains exactly 1 mol of carbon atoms). Molar mass expressed in g mol$^{-1}$ has the same numerical value as relative atomic mass ($A_r$), which is a unitless quantity:

$$M(^{12}C) = 12 \text{ g mol}^{-1}$$
$$A_r(^{12}C) = 12$$

For molecular species, we use the term *relative molecular mass* ($M_r$) instead of relative atomic mass ($A_r$). The $M_r$ values are also unitless. To find the relative molecular mass of a compound, we need to add together the $A_r$ values of all atoms in this compound.

🔑 **Key term**

The **molar mass** (*M*) of a chemical substance in g mol$^{-1}$ is the mass of 1 mol of that substance.

---

### Worked example: Calculating the relative molecular mass

**5.** The formula of water is $H_2O$. Calculate its relative molecular mass and molar mass.

*Solution*

The molecule contains two hydrogen atoms and one oxygen atom.
From table 1, $A_r(H) = 1.01$ and $A_r(O) = 16.00$, so:

$$M_r(H_2O) = 2 \times 1.01 + 16.00 = 18.02.$$

Since $M_r$ and $M$ have the same numerical value, $M(H_2O) = 18.02$ g mol$^{-1}$.

---

## Question

11 Calculate relative molecular masses and molar masses of the following substances:
   a) elementary chlorine, $Cl_2$; b) ammonia, $NH_3$; c) sulfuric acid, $H_2SO_4$.
   The $A_r$ values of elements are given in table 1 and the periodic table.

## Maths skills: Significant figures

Any measurement involves uncertainty, so the result of the measurement always has a limited number of *significant figures*. The length of a small object measured with a ruler typically has no more than three significant figures (sf) as shown in figure 14.

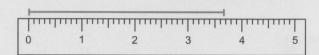

**Figure 14.** Measurement uncertainty when using a ruler

In this example, the measured length is 3.67 cm. The first two figures, 3 and 6, are certain, as the length is definitely greater than 3.6 but less than 3.7 cm. The last figure, 7, is uncertain, as the actual length could be 3.66 or 3.68 cm. There is absolutely no way of getting the fourth figure using this particular ruler, as we are not sure even about the third figure.

When the results of measurements are multiplied or divided, the value with the smallest number of sf determines the number of sf in the answer. For example, $3.67 \times 1.3 = 4.771 \approx 4.8$ (the answer must be rounded to two sf, as the second value has only two sf).

When the results of two or more measurements are added together or subtracted from one another, the value with the fewest decimal places determines the number of decimal places in the answer. For example, $3.67 + 1.3 = 4.97 \approx 5.0$ (the answer must be rounded to one decimal place, as the second value has only one decimal place).

In multi-step calculations, all intermediate results may be recorded with one or two extra significant figures, and only the final result should be rounded as explained above.

The amount, mass and molar mass of any substance are related as follows:

$$n = \frac{m}{M}$$

This is probably the most common expression in chemistry, as it is used in almost all stoichiometric calculations. Although the base SI unit of mass is kilogram, the masses of chemical substances are traditionally expressed in grams, and molar masses in g mol$^{-1}$.

 **Key term**

The **mole ratio** of elements in a chemical compound is the ratio of the amounts of these elements in the compound. For example, one mole of water ($H_2O$) contains two moles of hydrogen atoms and one mole of oxygen atoms, so the mole ratio of hydrogen to oxygen in water is 2:1.

**Question**

12 The relative atomic mass of lead is 207.20. Calculate the amount of lead in 1.0 g of this metal.

13 Calculate the mass of 15 mol of sulfuric acid, $H_2SO_4$.

The *mole ratio* can be used to calculate the percentage composition of a compound. (Unless stated otherwise, the term "percentage composition" refers to mass percentages.)

**Worked example: Calculating percentage composition**

**6.** Calculate the percentage composition of water.

*Solution*

Let $n(H_2O) = 1$ mol, then:

$$m(H) = 2 \text{ mol} \times 1.01 \text{ g mol}^{-1} = 2.02 \text{ g}$$

$$m(O) = 1 \text{ mol} \times 16.00 \text{ g mol}^{-1} = 16.00 \text{ g}$$

$$m(H_2O) = 1 \text{ mol} \times 18.02 \text{ g mol}^{-1} = 18.02 \text{ g}$$

$$\omega(H) = \frac{2.02 \text{ g}}{18.02 \text{ g}} \times 100\% \approx 11.2\%;$$

$$\omega(O) = 100\% - 11.2\% = 88.8\%.$$

**Question**

14 Calculate the percent composition of sulfuric acid, $H_2SO_4$.

In practice, chemists more often face the opposite problem of deducing the formula of an unknown compound from its percentage composition. The mass percentages of elements in a sample can be determined by various analytical techniques, such as fully automated combustion elemental analysis. In a typical experiment, the sample is burned in excess oxygen, and the volatile combustion products are trapped and weighed. These weights are then converted into mass percentages of chemical elements in the original sample.

**Worked example: Deducing the empirical formula**

**7.** Deduce the formula of a compound that contains 3.05% hydrogen and 96.95% sulfur by mass.

*Solution*

First of all, we need to check whether the mass percentages add up to 100%:

$$3.05\% + 96.95\% = 100\%$$

Therefore the compound contains no other elements, and its formula can be represented as $H_xS_y$.

Let's assume that $m(H_xS_y) = 100$ g, so the mass percentages of elements become numerically equal to their masses:

$$m(H) = \frac{100 \text{ g} \times 3.05\%}{100\%} = 3.05 \text{ g}$$

$$m(S) = 100 \text{ g} - 3.05 \text{ g} = 96.95 \text{ g}$$

$$n(H) = \frac{3.05 \text{ g}}{1.01 \text{ g mol}^{-1}} \approx 3.02 \text{ mol}$$

$$n(S) = \frac{96.95 \text{ g}}{32.07 \text{ g mol}^{-1}} \approx 3.02 \text{ mol}$$

$$n(H):n(S) = 3.02 \text{ mol}:3.02 \text{ mol} = 1:1.$$

Therefore, the *empirical formula* of this compound is HS.

**Key term**

The **empirical formula** is the simplest integer ratio of elements present in the compound.

Empirical formulae, like the one found in worked example 7, are determined from experimental data. An empirical formula of a specific compound might or might not be the same as its molecular formula, which shows the exact number of atoms of each element in the molecule (see *1.2 Chemical substances, formulae and equations*). In the worked example, the empirical formula HS seems unlikely, as we know that hydrogen is monovalent and sulfur is normally divalent, so the valences of these two elements do not match.

We can try to guess the correct molecular formula by multiplying the number of each atom in the empirical formula by the same factor (2, 3, and so on). For example, doubling the formula HS will produce $H_2S_2$, which corresponds to the following structure:

$$\underset{H}{\overset{}{\diagup}} S \diagdown \underset{S}{\overset{}{\diagdown}} \diagup \overset{H}{\overset{}{\diagup}}$$

This compound, hydrogen disulfide, does exist, so the guess is probably correct. But we can never be sure about it without additional experimental evidence. Such evidence can be obtained by determining the molar mass of this compound.

**Question**

15 Deduce the empirical formula of a compound that contains 30.45 % nitrogen and 69.55 % oxygen by mass.

**Key term**

**Avogadro's Law** states that equal volumes of any two gases at the same temperature and pressure contain equal numbers of molecules.

**Molar volume** $(V_m)$ of an ideal gas is equal to 22.7 $dm^3$ $mol^{-1}$ at **standard temperature and pressure (STP)** ($T = 273$ K (0°C) and $p = 100$ kPa).

This in turn gives us the **ideal gas law**, $pV = \dfrac{m}{M} \times RT$.

### Calculations involving gases

In 1811, Amedeo Avogadro suggested that equal volumes of any two gases at the same temperature and pressure contain equal numbers of molecules. This hypothesis has been confirmed in many experiments and is now known as *Avogadro's Law*. An important consequence of Avogadro's Law is that under the same conditions, the mass of a certain volume of a gas or vapour is proportional to the molar mass of this gas:

$$m = \frac{V}{V_m} \times M$$

In this expression, $V_m$ is the *molar volume* of an ideal gas, which is equal to 22.7 $dm^3$ $mol^{-1}$ at $T = 273$ K (0°C) and $p = 100$ kPa (figure 15). These conditions, known as *standard temperature and pressure (STP)*, are commonly used for comparing and determining the properties of gases. Although the molar volumes of real gases might differ slightly from 22.7 $dm^3$ $mol^{-1}$, these differences are very small, so in this book we will assume that the standard $V_m$ value can be applied to any gas.

If we know the mass and volume of a gas sample at STP, we can rearrange the above expression to find the molar mass of the gas:

$$M = \frac{m}{V} \times V_m$$

The ratio $m/V$ is the gas *density* $(\rho)$, so the molar mass of a gas and its density are related as follows:

$$M = \rho \times V_m$$

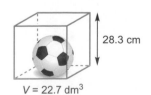

28.3 cm

$V = 22.7$ $dm^3$

**Figure 15.** Molar volume of an ideal gas compared with a soccer ball

For example, if the unknown gas with the empirical formula HS has a density of 2.91 g dm$^{-3}$ at STP, then:

$$M = 2.91 \text{ g dm}^{-3} \times 22.7 \text{ dm}^3 \text{ mol}^{-1} \approx 66.1 \text{ g mol}^{-1}$$

This value is very close to the molar mass of $H_2S_2$ (66.16 g mol$^{-1}$), which confirms our earlier suggestion.

**Question**

16 Calculate the molar mass of a gaseous oxide of sulfur and deduce its molecular formula if 0.250 dm$^3$ of this oxide at STP has a mass of 0.706 g.

If a substance is liquid or solid at STP, its molar mass can be determined under different conditions (typically at a higher temperature) using the *ideal gas law*:

$$pV = \frac{m}{M} \times RT$$

The *universal gas constant* ($R$) in this expression has a value of 8.31 J mol$^{-1}$ K$^{-1}$. To obtain the molar mass in g mol$^{-1}$, it is convenient to measure the pressure ($p$) in kPa, volume ($V$) in dm$^3$, mass ($m$) in g and temperature ($T$) in K.

Temperatures in °C must be converted to kelvins: $T_K = T_{°C} + 273.15$.

**Question**

17 A sealed container with a volume of 0.400 dm$^3$ was charged with 2.32 g of an unknown liquid and heated to 353 K until all liquid was vaporized. At that moment, the pressure in the container reached 124 kPa.

a) Calculate the molar mass of the liquid.

b) Deduce the molecular formula of the liquid if it contains 22.55% phosphorus and 77.45% chlorine by mass.

### Calculations involving chemical equations

The stoichiometric coefficients in any chemical equation show the ratio between reactants and products. For example, the following equation represents the formation of water from molecular hydrogen and oxygen:

$$2H_2(g) + O_2(g) \rightarrow 2H_2O(l)$$

This equation tells us that each $O_2$ molecule reacts with two $H_2$ molecules, producing two molecules of water. In other words, the molecular ratio between $H_2$, $O_2$ and $H_2O$ is 2:1:2.

Since the amount of each substance is proportional to the number of its molecules (or other structural units), the mole ratio between the reactants and products will be the same as their molecular ratio. In our example, each mole of oxygen will react with two moles of hydrogen, producing two moles of water.

In turn, mole ratios can be used to calculate mass and volume ratios. Therefore, if we know the amount, mass or volume of any substance participating in a chemical reaction, we can find the amounts, masses and volumes of all other reactants and products.

**Worked example: Calculating amounts, masses and volumes**

**8.** Suppose that the above reaction produced 3.6 g of water. Calculate the amounts, masses and volumes of the reactants at STP.

*Solution*

$$M(H_2O) = 2 \times 1.01 + 16.00 = 18.02 \text{ g mol}^{-1}$$
$$n(H_2O) = 3.6 \text{ g}/18.02 \text{ g mol}^{-1} \approx 0.20 \text{ mol}$$
$$n(H_2) = n(H_2O) = 0.20 \text{ mol}$$
$$n(O_2) = n(H_2O)/2 = 0.20 \text{ mol}/2 = 0.10 \text{ mol}$$
$$m(H_2) = 2 \times 1.01 \text{ g mol}^{-1} \times 0.20 \text{ mol} \approx 0.40 \text{ g}$$
$$m(O_2) = 2 \times 16.00 \text{ g mol}^{-1} \times 0.10 \text{ mol} \approx 3.2 \text{ g}$$
$$V(H_2) \text{ at STP} = 0.20 \text{ mol} \times 22.7 \text{ dm}^3 \text{ mol}^{-1} \approx 4.5 \text{ dm}^3$$
$$V(O_2) \text{ at STP} = 0.10 \text{ mol} \times 22.7 \text{ dm}^3 \text{ mol}^{-1} \approx 2.3 \text{ dm}^3.$$

Note that we cannot use $V_m$ to calculate the volume of liquids, such as water in this example.

A single numerical value (in our case, the mass of the product) has allowed us to determine the quantities of all reacting species. This is the power of chemical equations: in just one line, they represent many qualitative and quantitative characteristics of the chemical change.

All calculations based on chemical equations begin from determining the amount of any species (reactant or product) present in the equation. Once that is done, other quantities can be calculated, as the amounts of all other reacting species are proportional to their stoichiometric coefficients. In turn, the mass of each substance and the volume of each gas can be found in one step by multiplying their amounts with $M_r$ and $V_m$ values, respectively.

**Question**

18 The combustion of hydrogen sulfide proceeds as follows:

$$2H_2S(g) + 3O_2(g) \rightarrow 2H_2O(l) + 2SO_2(g)$$

Calculate the volume of consumed oxygen and the masses of the reaction products if the volume of combusted hydrogen sulfide was 0.908 dm³. All volumes are measured at STP.

From Avogadro's law, it follows that the amount of a gaseous substance is proportional to its volume:

$$n = \frac{m}{M} = \frac{V}{V_m}$$

Therefore, the volumes of reacting species measured under the same conditions are proportional to the amounts of these species:

$$\frac{n_1}{n_2} = \frac{V_1}{V_2}$$

In turn, the amounts of reactants and products are proportional to their stoichiometric coefficients. As a result, if we know the volume of any gas consumed or produced in the reaction, the volumes of other gaseous substances can be found without calculating their amounts.

> ### Worked example: Calculating volume using stoichiometric coefficients
>
> 9. Deduce the volume of oxygen consumed in question 18.
>
> *Solution*
>
> $$V(O)_2 = \frac{3}{2} \times V(H_2S) = \frac{3}{2} \times 0.908 \text{ dm}^3 \approx 1.36 \text{ dm}^3$$

## The limiting reactant

So far, we have assumed that all reactants are used up completely in the course of the reaction, and the final mixture contains only the reaction products. In real life, chemical reactions rarely go to completion, so some starting materials are often left over. In addition, one of the reactants may have been used in excess, so some of it will remain in the final mixture even if all other reactants are fully consumed.

For example, suppose that we mixed 1.0 mol of hydrogen with 3.0 mol of oxygen and ignited the mixture. As usual, hydrogen and oxygen will react in a 2:1 ratio, producing water:

$$2H_2(g) + O_2(g) \rightarrow 2H_2O(l)$$

The reaction will stop when all the hydrogen is consumed. If 1.0 mol of hydrogen has reacted with 0.5 mol of oxygen, another 3.0 – 0.5 = 2.5 mol of oxygen remains unreacted. Therefore, the final mixture will contain 1.0 mol of water and 2.5 mol of oxygen.

It is helpful to record the amounts of all substances as follows:

$$2H_2(g) + O_2(g) \rightarrow 2H_2O(l)$$

| | $2H_2(g)$ | $+ O_2(g)$ | $\rightarrow 2H_2O(l)$ |
|---|---|---|---|
| $n_{init}$ / mol | 1.0 | 3.0 | 0 |
| $\Delta n$ / mol | −1.0 | −0.5 | + 1.0 |
| $n_{fin}$ / mol | 0 | 2.5 | 1.0 |

The first row under the equation represents the initial mixture, where the substances can be present in any proportions. Indeed, we can take any amounts of reactants and mix them together, regardless of stoichiometric coefficients.

The second row shows how the amount of each substance changes in the course of the reaction. The sign before each amount shows whether it decreases (for reactants) or increases (for products). These changes must be proportional to stoichiometric coefficients, as all substances are consumed or produced according to the equation.

The last row represents the final mixture, where the amounts of all substances are calculated as $n_{fin} = n_{init} + \Delta n$. As with the initial mixture, the composition of the final mixture is not related to stoichiometric coefficients.

In this example, the amount of the reaction product (water) is limited by the amount of hydrogen, which is completely consumed in the reaction. In other words, hydrogen is the *limiting reactant*, while oxygen is present in excess.

 **Key term**

A **limiting reactant** is the reactant used up completely while other reactants are present in excess. It controls the amount of product formed in the reaction.

The concept of the limiting reactant is very important, as it allows us to determine the extent of reaction. For example, if we double the amount of hydrogen, the amount of water formed will also double:

$$2H_2(g) + O_2(g) \rightarrow 2H_2O(l)$$

| | $2H_2(g)$ | $O_2(g)$ | $2H_2O(l)$ |
|---|---|---|---|
| $n_{init}$ / mol | 2.0 | 3.0 | 0 |
| $\Delta n$ / mol | −2.0 | −1.0 | +2.0 |
| $n_{fin}$ / mol | 0 | 2.0 | 2.0 |

However, if we double the amount of oxygen (which was already in excess), the amount of water produced in the reaction will not change at all:

$$2H_2(g) + O_2(g) \rightarrow 2H_2O(l)$$

| | $2H_2(g)$ | $O_2(g)$ | $2H_2O(l)$ |
|---|---|---|---|
| $n_{init}$ / mol | 1.0 | 6.0 | 0 |
| $\Delta n$ / mol | −1.0 | −0.5 | +1.0 |
| $n_{fin}$ / mol | 0 | 5.5 | 1.0 |

Therefore in this case, the larger the excess of oxygen we use, the greater amount of it will remain unreacted.

## Question

**Q**

19 Hydrogen and chlorine react with each other to produce hydrogen chloride:

$$H_2(g) + Cl_2(g) \rightarrow 2HCl(g)$$

A mixture of 4.54 dm³ of hydrogen and 2.27 dm³ of chlorine was heated until the reaction was complete. Calculate the volumes of each substance in the final mixture. All volumes are measured at STP.

## The reaction yield

In any chemical reaction, the total mass of the reaction products is equal to the total mass of the consumed reactants. This principle, known as the *law of conservation of mass*, follows from the atomic theory. Since atoms cannot be created or destroyed, their total number and mass cannot be affected by chemical changes.

**Figure 16.** Top: Mikhail Lomonosov (1711–1765); bottom: Antoine Lavoisier (1743–1794)

**Question**

20 Calculate the masses of reactants and products in question 19 and verify that the law of conservation of mass is valid for the reaction between hydrogen and chlorine.

21 Magnesium metal (10.21 g) was burned in excess oxygen to produce magnesium oxide. Determine the initial volume of oxygen at STP if the volume of oxygen left after the reaction was 6.81 dm³ at STP.

When chemical reactions are carried out in a laboratory, the actual amounts of the reaction products are usually lower than the amounts predicted by the equation. This can happen for many reasons, including incomplete conversion of the reactants or simply because some product was lost during its isolation and purification. In such cases, we can calculate the *reaction yield*, which is the ratio of the practical and theoretical amounts of the product. Since the amount of an individual substance is proportional to its mass ($n = m/M$), the yield can also be found as the ratio of the practical and theoretical masses of the product:

 **Key term**

**Reaction yield** is the ratio of the practical and theoretical amounts of the product.

$$Yield = \frac{n_{pract}}{n_{theor}} \times 100\% = \frac{m_{pract}}{m_{theor}} \times 100\%$$

**Worked example: Calculating the percentage yield**

**10.** Suppose that we mixed 10.0 g of calcium metal and 9.62 g of sulfur, heated the mixture for some time, and collected 17.0 g of calcium sulfide, CaS. Calculate the yield of calcium sulfide.

*Solution*

First we need to find the amounts of all substances:

$$n(Ca) = 10.0 \text{ g}/40.08 \text{ g mol}^{-1} \approx 0.250 \text{ mol}$$
$$n(S) = 9.62 \text{ g}/32.07 \text{ g mol}^{-1} \approx 0.300 \text{ mol}$$
$$n(CaS) = 17.0 \text{ g}/72.15 \text{ g mol}^{-1} \approx 0.236 \text{ mol}.$$

According to the equation, the limiting reactant is calcium, as the two elements are consumed in a 1:1 ratio:

$$Ca(s) + S(s) \rightarrow CaS(s)$$

The theoretical amount of calcium sulfide is 0.250 mol (the same as the initial amount of the limiting reactant, calcium). However, the reaction produced only 0.236 mol of this compound, so the yield of calcium sulfide is:

$$(0.236 \text{ mol}/0.250 \text{ mol}) \times 100\% = 94.4\%.$$

The same result could be obtained through practical and theoretical masses of calcium sulfide:

$$m(CaS)_{theor} = 0.250 \text{ mol} \times 72.15 \text{ g mol}^{-1} \approx 18.0 \text{ g}$$
$$Yield = (17.0 \text{ g}/18.0 \text{ g}) \times 100\% \approx 94.4\%.$$

**Question**

22 Calculate the percentage yield of the reaction between 9.443 g of aluminium and 7.945 dm³ (STP) of oxygen that produced 17.13 g of aluminium oxide.

The yields of chemical reactions are particularly important in the chemical industry, where the loss of a tiny percentage of the final product could mean a significant drop in profit. At the same time, low reaction yields increase the amount of waste, which needs to be disposed of safely or reused. The development of highly efficient synthetic procedures with low environmental impact is reflected in the concept of *green chemistry*, which is now adopted by the majority of commercial and research organizations around the world.

> **DP ready**    **Nature of science**
>
> Green chemistry
>
> The term "green chemistry" was coined in 1991 by American chemists Paul Anastas and John Warner, who formulated 12 principles for their approach to chemical technology. These principles include the use of nonhazardous chemicals and solvents, reduction of energy consumption and waste production ("the best form of waste disposal is not to create it in the first place"), choice of renewable materials, and prevention of accidents. The philosophy of green chemistry has passed into national and international laws, which restrict the use of certain chemical substances and promote environmentally friendly technologies.

 **Internal link**

Later you will see how the laws of stoichiometry can be applied to redox processes (3.3 Redox processes), electrolytes (4.2 Concentration expressions and stoichiometry), chemical energetics (5.1 Thermochemistry), kinetics (5.2 Chemical kinetics), equilibria (5.3 Chemical equilibrium) and acid–base reactions (6.3 The concept of pH).

## Stoichiometry—a general approach

Solving a stoichiometry problem is often a challenge that requires handling a variety of data and performing multi-stage mathematical calculations. Although there is no single strategy, most problems can be approached as follows.

1. Write and balance the chemical equation(s) for any chemical changes mentioned in the problem.

2. Calculate the amounts of as many substances as you can using the formulae $n = m/M$ and $n = V/V_m$. If you know a mass percentage, convert it first to mass (if possible) and then to amount.

3. Write all known amounts of substances below their formulae in the chemical equation. Any missing values may suggest the next step of the solution.

4. Determine the limiting reactant and use stoichiometric coefficients to calculate the amount changes for all substances. Remember that these coefficients do not reflect the initial or final amounts of reactants and products.

5. Check the mass balance. The total mass of products must be equal to the total mass of reactants. If it is not the case, revise your solution.

6. Check that the answer makes sense. All percentages should add up to 100%, and no individual percentage can be greater than 100%. The yield of the final product must not exceed 100%. Finally, treat very low or very high percentages with caution.

These simple rules will help you to avoid many common errors and solve the problem in the fewest steps. We will continue the discussion of stoichiometric relationships in the next chapters of this book.

# Chapter summary

In this chapter, you have learned about the atomic theory, chemical and physical changes, chemical formulae, equations and stoichiometry.

Before moving further, check that you have a working knowledge of the following concepts and definitions.

- ☐ All matter consists of atoms, which cannot be created or destroyed by chemical changes.
- ☐ The atom is the smallest unit of a chemical element.
- ☐ An atom consists of a small, dense, positively charged nucleus surrounded by negatively charged electrons.
- ☐ Atoms can form ions by losing or gaining electrons, or combine together into molecules by sharing electrons.
- ☐ Elementary substances contain atoms of one element while compounds contain atoms of two or more elements.
- ☐ Pure substances have definite compositions that can be represented by chemical formulae.
- ☐ An empirical formula shows the simplest integer ratio of atoms in a compound.
- ☐ A molecular formula shows the actual number of atoms in the molecule.
- ☐ One mole of any substance contains $6.02 \times 10^{23}$ structural units.
- ☐ The amount, mass and molar mass of a substance are related by the expression $n = \dfrac{m}{M}$.
- ☐ One mole of any gas has a volume of 22.7 dm³ at STP (273 K and 100 kPa).
- ☐ The amount and volume of a gas are related by the expression $n = \dfrac{V}{V_m}$.
- ☐ Chemical equations represent chemical changes where the reactants are transformed into the products.
- ☐ Stoichiometric coefficients show the mole ratio of the substances consumed or produced in the course of the reaction.
- ☐ The total mass of the reaction products is equal to the total mass of the reactants.
- ☐ The reaction yield is the ratio between the practical and theoretical amounts of the product.

## Additional problems

1. Ammonia ($NH_3$) is a colourless, strong-smelling, toxic gas that can be produced from hydrogen and nitrogen. Under a pressure of 1 MPa, ammonia condenses into a liquid at room temperature. Evaporation of liquid ammonia absorbs a significant amount of heat, which makes this compound a common refrigerant. Ammonia is flammable and typically produces nitrogen gas and water when burned in air. Identify all physical and chemical changes described in this paragraph and represent these changes using equations and state symbols.

2. Dalton's atomic theory consisted of five postulates, as explained in the text. Examine each of these postulates and suggest whether it agrees with the modern views on the atomic structure or not. Support your answer by specific examples.

3. Tritium is a radioactive isotope of hydrogen with a mass number of 3. State the number of protons, neutrons and electrons in an atom of tritium.

4. Naturally occurring sulfur has four isotopes with the following atomic percentages: $^{32}S$ (95.02%), $^{33}S$ (0.75%), $^{34}S$ (4.21%) and $^{36}S$ (0.02%). Calculate the average $A_r$ value for sulfur. Suggest why your result differs slightly from the $A_r$ value given in table 1.

5. Hydrogen peroxide ($H_2O_2$) and hydrazine ($N_2H_4$) are covalent molecules that contain single O–O and N–N bonds, respectively. Many molecules with such bonds decompose easily, producing large amounts of heat. Because of that, both hydrogen peroxide and hydrazine are used as fuel in rocket engines.

   a) Draw the structural formulae of hydrogen peroxide and hydrazine, representing covalent bonds with lines.

   b) Decomposition of hydrogen peroxide produces water and an elementary substance while decomposition of hydrazine produces ammonia ($NH_3$) and another elementary substance. Deduce and balance chemical equations for these processes.

   c) Suggest the most likely state symbols for all reaction products when liquid hydrogen peroxide and liquid hydrazine are used as rocket fuels.

6. To visualize the mole, a chemistry student decided to pile up $6.02 \times 10^{23}$ grains of sand. Estimate the time needed to complete this project if an average grain of sand weighs 5 mg, and the student can shovel 50 kg of sand per minute.

7. Calculate the percentage composition of magnesium hydroxide, $Mg(OH)_2$.

8. Calculate the density, in $g\ dm^{-3}$, of chlorine gas at: a) STP; b) 100°C and 30.0 kPa.

9. Carbon forms several gaseous compounds with fluorine. Deduce empirical, molecular and structural formulae for three of these compounds using the mass percentages and densities from the table below.

| Compound | $\omega(C)$ / mass % | $\rho$ at STP / g dm$^{-3}$ |
|---|---|---|
| X | 13.65 | 3.88 |
| Y | 24.02 | 4.41 |
| Z | 17.40 | 6.08 |

10. A reaction between hydrogen sulfide and sulfur dioxide proceeds as follows:

$$2H_2S(g) + SO_2(g) \rightarrow 3S(s) + 2H_2O(l)$$

   Deduce the limiting reactant and determine the molar composition of the final mixture if the initial mixture contained 11.35 dm³ of hydrogen sulfide and 18.16 dm³ of sulfur dioxide at STP.

11. Magnesium metal (100.0 g) was heated with an equal mass of elemental phosphorus until the reaction was complete. Determine the mass percentages of all substances in the final mixture.

12. Calcium carbonate ($CaCO_3$, limestone) and calcium oxide (CaO, quicklime) are widely used in construction and steelmaking. Above 825°C, calcium carbonate decomposes into calcium oxide and carbon dioxide, $CO_2$. Calculate the percentage yield of this reaction if the decomposition of 20.0 kg of calcium carbonate in an industrial furnace released 3.86 m³ (STP) of carbon dioxide.

# Electron structure and chemical bonding

> " As Bohr has pointed out, the properties of gravitation and radioactivity, which are entirely uninfluenced by chemical or physical agencies, must be ascribed mainly if not entirely to the nucleus, while the ordinary physical and chemical properties are determined by the number and distribution of the external electrons. "

**Ernest Rutherford, *The Structure of the Atom* (1914)**

## Chapter context

In the previous chapter, we discussed the atomic structure of matter and the fact that atoms combine in definite proportions, forming chemical bonds by losing, gaining or sharing their electrons. However, we did not explain why atoms of each chemical element have a specific **valence** and use only a certain number of their electrons for bonding. We can now solve this mystery by examining the **electron structure** of atoms and the ways this structure defines the **properties** of **chemical elements**.

## Learning objectives

In this chapter you will learn about:

→ the **quantum model** of the atom

→ the concept of **electron configuration**

→ how the **atomic number** influences **chemical properties**

→ how the **periodic table** is derived from patterns known as the **periodic law**

→ using the periodic table to deduce **chemical structures**.

## 🔑 Key terms introduced

→ Photons
→ Quantized states and the principal quantum number $(n)$
→ Energy levels/shells; ground states and excited states
→ Atomic orbitals, energy sublevels/subshells
→ The Aufbau principle
→ Valence electrons and core electrons
→ Periods and groups and the periodic law
→ Atomic radius $(r)$
→ Ionization energy $(E_i)$ and electron affinity $(E_{ea})$
→ Electronegativity $(\chi)$
→ Intermolecular forces; hydrogen bonds, van der Waals forces, electric dipoles
→ Coordinate bonds
→ Valence shell electron pair repulsion (VSEPR) theory
→ Covalent network solids; giant covalent structures
→ Simple molecular structures
→ Allotropes

## 2.1 Electron configuration

### The quantum atom

The Rutherford model described the atom as a planetary system with a small, dense, positively charged nucleus surrounded by negatively charged electrons of negligible mass. Although this model could explain many experimental observations, it had serious theoretical problems. In particular, classical electrodynamics predicted that orbiting electrons would radiate energy and quickly fall into the nucleus, making any prolonged existence of atoms impossible. Another problem was related to the light radiated by excited atoms, which consisted of narrow coloured lines instead of a continuous spectrum (figure 1).

## ⊗⊗ Internal link

The Rutherford model of the atom is described in **1.1 The particulate nature of matter**.

violet  blue      green                                    red

**Figure 1.** Emission spectrum of atomic hydrogen

The first attempt to overcome these problems was made in 1913 by the Danish physicist Niels Bohr. He suggested that electrons could occupy only certain "stationary" orbits around the nucleus, and did not radiate

**Figure 2.** Niels Bohr (1885–1962)

energy when staying in those orbits. Each orbit was associated with a specific *energy level*: the larger the orbit, the greater the energy. An electron could jump from a lower to a higher orbit by absorbing a *photon* of the right amount of energy to make up the difference. Similarly, an electron falling from a higher to a lower orbit would emit a photon with exactly the same energy.

Since electrons in the Bohr model of the atom could have only certain, well-defined energies, their transitions between stationary orbits could absorb or emit photons of specific wavelengths, producing characteristic lines in the atomic spectra. By measuring the wavelengths of these lines, it was possible to calculate the energies of electrons in stationary orbits. For a hydrogen atom, the electron energy ($E_n$) in joules could be related to the energy level number ($n$) by a simple equation:

$$E_n = \frac{2.176 \times 10^{-18}}{n^2}$$

This equation clearly represents the *quantum* nature of the atom, where the electron energy can have only discrete, *quantized* values. These values are characterized by integer or half-integer parameters, known as *quantum numbers*. The *principal quantum number* ($n$) can take only positive integer values (1, 2, 3 and so on), where greater numbers mean higher energy. The most stable state of the hydrogen atom is the state at $n = 1$, where the electron has the lowest possible energy. This energy level is known as the *ground state* of the atom. In contrast, the energy levels with $n = 2$, 3 and so on are called *excited states*. Atoms in excited states are unstable and spontaneously return to the ground state by emitting photons of specific wavelengths.

### Atomic orbitals

Although the Bohr model could predict the spectra of the hydrogen atom and some ions with a single electron, it had many limitations and could not be applied to multi-electron species. Very soon it became clear that electrons in an atom do not travel around the nucleus like planets around the Sun but rather surround the nucleus as diffuse clouds of various sizes and shapes. In other words, electrons do not follow specific paths within the atom but occupy its entire volume. By doing so, they behave not as particles of definite sizes but rather as three-dimensional waves oscillating around the nucleus. To distinguish these diffuse, wave-like clouds from orbits, they are called *atomic orbitals*.

According to modern quantum mechanics, the electron in a hydrogen atom in the ground state ($n = 1$) forms a spherical cloud (figure 3). This type of atomic orbital is designated by the letter s. In higher energy levels ($n > 1$), the electron clouds of s orbitals increase in size and form alternating layers of high and low density.

**Key term**

A **photon** is a particle of light, or electromagnetic energy.

**Key term**

**Quantized** states can only have certain values, with no possible states in between.

**Quantum numbers** are used to characterize these discrete values. The **principal quantum number (n)** of an electron tells us which **energy level** the electron is occupying in an atom.

The **ground state** of the atom is its most stable state, in which all its electrons have the lowest possible energies; in **excited states** the electrons have more energy and the atom has an unstable electron configuration.

**Key term**

Energy levels in an atom are often referred to as **electron shells**. Similarly, each set of **atomic orbitals** of the same energy is called an **energy sublevel** or **electron subshell**. In this book, we will be using only the terms "electron level" and "electron sublevel".

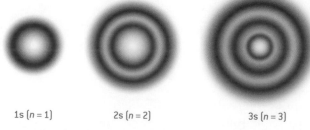

1s ($n = 1$)     2s ($n = 2$)     3s ($n = 3$)

**Figure 3.** Electron clouds of s orbitals with various principal quantum numbers ($n$)

Experimental data and mathematical calculations show that each energy level (electron shell) can accommodate up to $2n^2$ electrons. The first level ($n = 1$) consists of a single atomic orbital that can hold up to two electrons with opposite *spin quantum numbers*, or simply *spins* ($m_s$). The electron spin has only two possible values, $+\frac{1}{2}$ and $-\frac{1}{2}$, and characterizes certain magnetic properties of the electron.

## Question

1   Calculate the maximum numbers of electrons in the second and third energy levels.

### DP ready    Theory of knowledge

**Electron spin** is often interpreted as the rotation of the electron around its own axis. However, this interpretation has no physical basis: electrons in atoms behave like waves, and a wave cannot rotate. Unfortunately, neither the spin nor the wave-like behaviour of electrons can be visualized in any way, as they have no analogues in our everyday life and can be expressed only in mathematical form. This lack of visualization does not undermine the quantum theory but rather shows the limits of human perception and, at the same time, the power of mathematics as the language of science.

The second energy level consists of one spherical and three dumbbell-shaped orbitals, which are designated 2s and 2p, respectively (figure 4). Each 2p orbital consists of two lobes pointing away from the atomic nucleus. The axes of p orbitals are perpendicular to one another, which reduces the repulsion between the electrons occupying these orbitals.

The orbital type is defined by the *azimuthal quantum number*, $\ell$, which can take non-negative values from 0 to $n - 1$. For the first energy level ($n = 1$), the only possible value of $\ell$ is 0, which corresponds to the spherical 1s orbital. For the second energy level ($n = 2$), the possible $\ell$ values are 0 and 1, which correspond to spherical 2s and dumbbell-shaped 2p orbitals, respectively. Traditionally, azimuthal quantum numbers are represented by letters, as shown in table 1.

**Table 1.** Azimuthal quantum numbers ($\ell$)

| $\ell$ | 0 | 1 | 2 | 3 | 4 |
|---|---|---|---|---|---|
| **Orbital type** | s | p | d | f | g |
| **Orbital shape** | sphere | dumbbell | | various | |

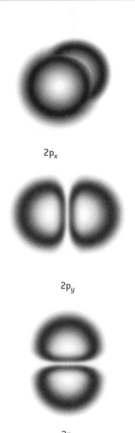

The orientation of a particular p orbital is defined by its *magnetic quantum number*, $m_\ell$, which can take integer values from $-\ell$ to $+\ell$, including zero. Since all 2p orbitals have the same azimuthal quantum number ($\ell = 1$), the three possible values of $m_\ell$ for these orbitals are $-1$, 0 and $+1$.

2p$_x$

2p$_y$

2p$_z$

**Figure 4.** Electron clouds of 2p orbitals; the subscript indices refer to the orbital orientation along the three Cartesian axes $x$, $y$ and $z$

**Figure 5.** Energy levels in atoms resemble ladders with varying distances between the rungs. Electrons cannot exist between energy levels, much like how you cannot stand between the rungs of a ladder

According to the *Pauli exclusion principle*, any two electrons in an atom must differ in at least one quantum number. Therefore, the combination of the four quantum numbers ($n$, $\ell$, $m_\ell$ and $m_s$) fully defines the state of any electron in an atom. All possible combinations of $n$, $\ell$ and $m_\ell$ for the first three energy levels are listed in table 2.

**Table 2.** Quantum numbers and atomic orbitals for $n = 1$ to 3; each orbital can contain two electrons, one with spin $+\frac{1}{2}$ and one with spin $-\frac{1}{2}$

| $n$ | $\ell$ | $m_\ell$ | Orbital | |
|---|---|---|---|---|
| | | | Symbol | Quantity |
| 1 | 0 | 0 | 1s | 1 |
| 2 | 0 | 0 | 2s | 1 |
| | 1 | $-1, 0, +1$ | 2p | 3 |
| 3 | 0 | 0 | 3s | 1 |
| | 1 | $-1, 0, +1$ | 3p | 3 |
| | 2 | $-2, -1, 0, +1, +2$ | 3d | 5 |

The shapes and orientation of atomic orbitals affect the direction of covalent bonds formed by the atom and, as a result, the structures of covalent molecules. We will discuss these concepts in more detail in *2.3 Chemical bonding* and *7.1 Structures of organic molecules*.

**Question**

2   Examine the relationships between quantum numbers $n$, $\ell$ and $m_\ell$ and expand table 2 to the next energy level ($n = 4$).

### Electron configurations

Electron configurations of atoms are often represented as orbital diagrams, in which atomic orbitals are shown as boxes and electrons as half-arrows. An upward half-arrow represents an electron with a positive spin ($m_s = +\frac{1}{2}$) while a half-arrow pointing down corresponds to an electron with $m_s = -\frac{1}{2}$. For example, the electron configuration of a hydrogen atom in the ground state can be represented as follows:

↑ hydrogen (H)
1s¹

The shorthand notation "1s¹" shows the energy level (first number), the orbital type (letter in the middle) and the number of electrons in that orbital (superscript number).

In the atom of the second lightest element, helium (He), two electrons occupy the same atomic orbital, so the electron configuration of helium in the ground state is 1s²:

↑↓ helium (He)
1s²

Since helium contains the maximum possible number of electrons for the first energy level, it is said to have a *complete* or *closed electron configuration*. Elements with closed electron configurations are extremely unreactive and are therefore known as *noble gases*. As suggested by their names, all these elements are gaseous under normal conditions and occur naturally only in the form of individual atoms.

The next element, lithium (Li), contains three electrons, only two of which can occupy the 1s orbital. The third electron goes into the second energy level ($n = 2$), which consists of one s and three p orbitals. The energy of the 2s orbital is slightly lower than the energies of the 2p orbitals, so the most stable electron configuration for lithium is $1s^2 2s^1$:

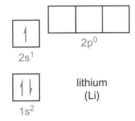

lithium
(Li)

In the last diagram, the boxes at the top represent atomic orbitals with the highest energy. Electrons fill the lowest available orbitals first, just as water rising in a pool would cover the steps of a ladder. This is called the *Aufbau principle*. For both electrons and water, this behaviour minimizes the overall potential energy of the system.

**Key term**

The **Aufbau principle** (from German "building up") states that in an atom's ground state electrons fill the lowest available orbitals first.

**Question**

3   The electron configuration of beryllium (Be) in the ground state is $1s^2 2s^2$. Draw the orbital diagram for this element.

An atom of boron (B) in the ground state has the electron configuration $1s^2 2s^2 2p^1$:

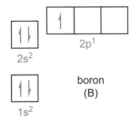

boron
(B)

The single 2p electron in boron can occupy any of the three orbitals, as they have equal energies. Traditionally, the half-arrow is drawn in the leftmost box, although it is a matter of personal preference.

The electron configuration of the next element, carbon (C), is $1s^2 2s^2 2p^2$. In this case, the two 2p electrons could potentially occupy either the same or two different orbitals.

## Key term

The number of electrons in an atom increases with the atomic number, so the electron configurations for heavier elements become lengthy. Such configurations are often written in condensed form, where complete energy levels are represented by the chemical symbols of noble gases. For example, the configuration of lithium ($1s^22s^1$) differs by a single 2s electron from the configuration of helium ($1s^2$), so it can be represented as $[He]2s^1$.

Condensed electron configurations emphasize the structure of outer energy levels of the elements. The electrons in these levels are known as **valence electrons**, as they can be transferred to or shared with other atoms to form chemical bonds. In contrast, **core electrons** in inner energy levels do not participate in chemical bonding, so they are not shown in condensed electron configurations.

## DP link

You will learn more about the electron structure of atoms by studying **2 Atomic structure** and **13 The periodic table — the transition metals.**

When two electrons stay in the same orbital, they have opposite spins and are said to be *paired*, in contrast to *unpaired* electrons in separate orbitals. Pairing of electrons requires energy, so it happens only when there are no available orbitals in the same sublevel. Therefore, the two 2p electrons in carbon remain unpaired and occupy different orbitals:

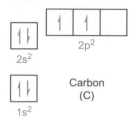

Note that both half-arrows point up, indicating that the two 2p electrons in carbon have the same spins ($m_s = +\frac{1}{2}$). This configuration corresponds to the minimum energy of the atom, so the spins of unpaired electrons in the ground state are always parallel to one another.

## Question

4 Atoms of nitrogen (N), oxygen (O), fluorine (F) and neon (Ne) contain three, four, five and six 2p electrons, respectively. Draw the orbital diagrams for these atoms in their ground states.

The third and fourth energy levels contain d and f orbitals (figure 6). These orbitals are typically filled after the s orbitals of the following levels, because they are higher in energy. As shown in figure 6, the 3d sublevel has a higher energy than 4s but lower than 4p, so 4s is filled with electrons first, followed by 3d and finally 4p. For the same reason, 4d orbitals are filled after 5s, and 4f orbitals are filled only after 6s. Generally, the following order is observed:

1s < 2s < 2p < 3s < 3p < 4s < 3d < 4p < 5s < 4d < 5p < 6s < 4f < 5d < 6p …

There are some exceptions to this rule, but we are not going to discuss them in this book.

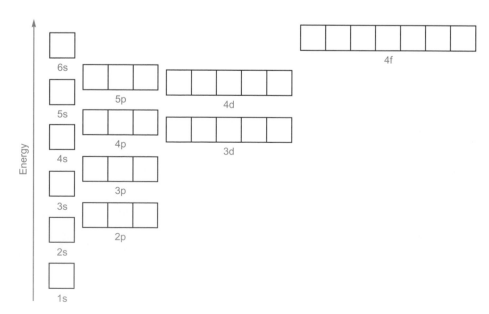

**Figure 6.** Orbital diagram for selected energy sublevels

**Question**

5   Copy the orbital diagram from figure 6 and complete it for the
    following elements in their ground states: **a)** aluminium (Al);
    **b)** chlorine (Cl); **c)** iron (Fe). Refer to table 1 from the previous
    chapter or the periodic table at the end of this book to determine
    the number of electrons in each atom.

So far, we have discussed only the electron configurations of neutral
atoms. To deduce the electron configuration of an ion, we can first
draw the configuration of a neutral atom and then remove or add
electrons according to the ionic charge. For example, we know that
the atomic number of sodium, Na, is 11, and this element forms singly
charged cations, Na$^+$. In a neutral atom of sodium, the number of
electrons is equal to its atomic number, so sodium has 11 electrons.
To produce the cation, we need to subtract one electron, so Na$^+$ has
11 − 1 = 10 electrons. Using the Aufbau principle, we can show that
the electron configuration of Na$^+$ is 1s$^2$2s$^2$2p$^6$, or [Ne]. In the following
diagrams, the configurations of Na and Na$^+$ are drawn side by side,
with 3d orbitals omitted to save space:

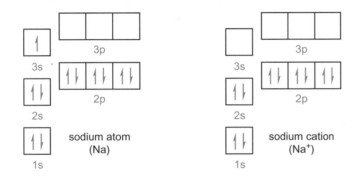

Sodium metal reacts violently with water and never occurs in nature.
In contrast, compounds containing sodium cations (such as table salt,
NaCl) are usually stable and unreactive. The reason for that is the
particular stability of electron configurations in which all energy levels
are either empty or completely filled with electrons. In the sodium
atom, the third energy level contains a single electron that is readily
lost in chemical reactions. The resulting sodium cation has a stable
electron configuration of a noble gas (Ne) with an empty third energy
level and a complete second energy level. As a result, sodium exists in
the form of Na$^+$ cations in all its compounds.

**Question**

6   Deduce the electron configuration of the Cl$^-$ anion and compare it with the configuration
    of the Cl atom from question 5. Explain why chlorine cannot form doubly charged anions.

## Key term

**Octet rule:** electron configurations with eight electrons in the highest occupied energy level are particularly stable

## The octet rule

In your answer to question 6, you probably noticed the similarity between electron configurations of $Na^+$ and $Cl^-$ ions. Indeed, both ions have eight electrons in their highest filled energy levels. The stability of such ions is reflected in the *octet rule*: atoms tend to lose, gain or share electrons in such a way that their outer energy levels contain eight electrons each. This rule has many exceptions, so you must use it with care. In particular, it does not work for hydrogen, helium, some elements with incompletely filled d orbitals, and atoms with a single electron in their outer p sublevel. However, in most other cases the octet rule provides a simple way to check whether an ion or molecule is likely to form in a chemical reaction.

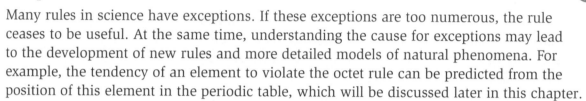

**DP ready** | **Nature of science**

### Exceptions as the rule

Many rules in science have exceptions. If these exceptions are too numerous, the rule ceases to be useful. At the same time, understanding the cause for exceptions may lead to the development of new rules and more detailed models of natural phenomena. For example, the tendency of an element to violate the octet rule can be predicted from the position of this element in the periodic table, which will be discussed later in this chapter.

In addition to the octet rule, it is important to note that atoms with one, two or three electrons in their outer energy level never form anions, as they would need to accept too many electrons to achieve a stable configuration. Instead, such atoms tend to lose outer electrons and form cations, leaving their highest energy level empty. Sodium, magnesium, aluminium and their cations ($Na^+$, $Mg^{2+}$ and $Al^{3+}$) are typical examples of such elements. The two exceptions are hydrogen and helium, with the electron configurations $1s^1$ and $1s^2$, respectively. Although hydrogen tends to lose its single electron and form the $H^+$ cation, it can also gain one electron to complete its outer level and form the $H^-$ anion with a very stable configuration $1s^2$. Helium has this configuration already, so it is completely unreactive and never forms chemical compounds.

## Question

7 Examine the electron configuration of aluminium from question 5 and suggest which electrons need to be removed to produce the $Al^{3+}$ ion. Explain why ions $Al^{2+}$ and $Al^{4+}$ are unstable.

Atoms with five, six and seven electrons in their outer energy level tend to form anions rather than cations, as they need to accept only a few electrons to satisfy the octet rule. Nitrogen, sulfur, chlorine and their anions ($N^{3-}$, $S^{2-}$ and $Cl^-$) are typical examples of such elements.

## Question

8 The neutral atom of selenium (Se) has the electron configuration $[Ar]3d^{10}4s^24p^4$. Suggest which ion(s) this element is likely to form in chemical reactions.

Atoms with four electrons in their outer energy level rarely form ions. Instead, they tend to share their electrons with other atoms and form covalent compounds. The most common example of such elements is carbon (C) with the electron configuration $1s^2 2s^2 2p^2$.

Noble gases, the elements with complete outer energy levels, already have stable electron configurations and thus are unreactive. Helium is the only noble gas with two electrons in the outer energy level; all other noble gases have electron configurations $ns^2 np^6$, where $n$ is the level number.

Now we know that the chemical properties of the elements and, to some extent, their compounds depend on the electron structure of their atoms. The electron structure of an atom is defined by a single parameter, the atomic number. In the next topic, we will reinforce this fundamental relationship between the atomic number and chemical properties by introducing the central concept of chemistry – the periodic law.

> **Internal link**
>
> The chemistry of carbon will be discussed in the last chapter of this book, **7 Organic chemistry**.

## 2.2 The periodic law

No chemistry textbook is complete without the periodic table. On a single chart, a considerable amount of our knowledge about the nature and properties of chemical elements is presented in a clear and elegant manner. For chemists, the periodic table is a powerful tool that can be used not only as a reference but also as a guide in their studies.

One of the first versions of the periodic table was created in 1869 by the Russian chemist Dmitri Mendeleev. He listed chemical elements in order of increasing atomic mass and showed that their properties recurred at regular intervals. This observation, known as the *periodic law*, allowed Mendeleev to develop a universal classification of all chemical elements, including those that had not been discovered at his time.

**Figure 7.** Dmitri Mendeleev (1834–1907)

### Periodicity

We can follow Mendeleev's reasoning by looking at the typical ionic charges and valences of the sixteen elements from lithium to argon:

| Element and $A_r$ | Li 6.94 | Be 9.01 | B 10.81 | C 12.01 | N 14.01 | O 16.00 | F 19.00 | Ne 20.18 | Na 22.99 | Mg 24.31 | Al 26.98 | Si 28.09 | P 30.97 | S 32.07 | Cl 35.45 | Ar 39.95 |
|---|---|---|---|---|---|---|---|---|---|---|---|---|---|---|---|---|
| Ionic charge (valence) | +1 | +2 | (3) | (4) | −3 | −2 | −1 | 0 | +1 | +2 | +3 | (4) | −3 | −2 | −1 | 0 |

The charges and valences of the first eight elements (lithium to neon) follow the same general pattern as the charges and valences of the next eight elements (sodium to argon). Therefore, we can split these elements into two *periods* and place each period in a separate row of the table. Now the elements with similar properties will appear in the same column, or *group*:

Groups

| Li 6.94 | Be 9.01 | B 10.81 | C 12.01 | N 14.01 | O 16.00 | F 19.00 | Ne 20.18 |
|---|---|---|---|---|---|---|---|
| Na 22.99 | Mg 24.31 | Al 26.98 | Si 28.09 | P 30.97 | S 32.07 | Cl 35.45 | Ar 39.95 |

Periods

Most periods in the periodic table begin with an active metal, such as lithium or sodium. The first period contains only two non-metallic elements, hydrogen and helium, which presents a problem. Helium, a typical noble gas, clearly belongs to the last group of the table. However, the position of hydrogen is ambiguous. Since hydrogen readily forms cations ($H^+$), it resembles lithium and sodium. On the other hand, hydrogen is a gas under normal conditions and can form anions ($H^-$), which makes it similar to fluorine and chlorine. Therefore, in older versions of the periodic table, hydrogen could appear in either of the two groups, with another group showing its symbol in parentheses:

| H | | | | | | (H) | He |
|---|---|---|---|---|---|---|---|
| 1.01 | | | | | | | 4.00 |
| Li | Be | B | C | N | O | F | Ne |
| 6.94 | 9.01 | 10.81 | 12.01 | 14.01 | 16.00 | 19.00 | 20.18 |
| Na | Mg | Al | Si | P | S | Cl | Ar |
| 22.99 | 24.31 | 26.98 | 28.09 | 30.97 | 32.07 | 35.45 | 39.95 |

In this book, we will show hydrogen in the first group only, as recommended by the International Union of Pure and Applied Chemistry (IUPAC).

## DP ready  Nature of science

### Describing the unknown

The periodic trends in chemical and physical properties of the elements were known, though not understood, before Dmitri Mendeleev developed his famous table. However, Mendeleev was the first to realise the fundamental nature of these trends and their predictive power. In particular, he pointed out that some atomic masses of the elements known at his time were determined incorrectly, and used the periodic law to estimate the correct values of these atomic masses. Mendeleev also left several gaps in his table for unknown elements and described their properties. For example, he suggested that a heavier analogue of silicon, now known as germanium, would be a grey solid with a relative atomic mass of 72 and a density of 5.5 g cm$^{-3}$. When germanium was discovered fifteen years later, its appearance (grey solid), relative atomic mass (72.63) and density (5.35 g cm$^{-3}$) closely matched Mendeleev's predictions, which was a triumph of the periodic law.

The predictive power of Mendeleev's periodic table illustrates the "risk-taking" nature of science and the importance of both inductive and deductive reasoning in making scientific claims.

Although Mendeleev arranged the elements according to their atomic masses, he understood that the periodic trends could be caused by unknown factors. Therefore, he used atomic masses as guidance only and changed the order of several elements so that they could fit into appropriate groups. For example, potassium ($A_r$ = 39.10) was placed into the table after argon ($A_r$ = 39.95), despite the fact that argon had a higher atomic mass. This put argon into the last group with another noble gas, neon, while potassium went into the next period and appeared below two other metals, lithium and sodium:

| H<br>1.01 | | | | | | | He<br>4.00 |
|---|---|---|---|---|---|---|---|
| Li<br>6.94 | Be<br>9.01 | B<br>10.81 | C<br>12.01 | N<br>14.01 | O<br>16.00 | F<br>19.00 | Ne<br>20.18 |
| Na<br>22.99 | Mg<br>24.31 | Al<br>26.98 | Si<br>28.09 | P<br>30.97 | S<br>32.07 | Cl<br>35.45 | Ar<br>39.95 |
| K<br>39.10 | | | | | | | |

This and many other inconsistencies in the periodic table were eliminated in 1913, when the British physicist Henry Moseley introduced the concept of atomic numbers and proposed the modern statement of the periodic law:

 **Internal link**

The atomic number was introduced in **1.1 The particulate nature of matter**.

 **Key term**

**Periodic law**: the properties of chemical elements are periodic functions of their atomic numbers.

Moseley's definition of periodicity provided a simple and unambiguous way of ordering the elements. For example, since the atomic number of argon was 18, it had to be placed in the periodic table before potassium with an atomic number of 19, regardless of relative atomic masses. (The higher relative atomic mass of argon is caused by its isotopic composition, as naturally occurring argon has a greater percentage of heavier isotopes than potassium does.)

**Figure 8.** Henry Moseley (1887–1915)

### The periodic table and electron configuration

Further evolution of the periodic law was closely related to the development of the quantum theory and the concept of electron configuration. By the mid-20th century, it became clear that each period in the table represented the filling of certain energy levels with electrons. In particular, the number of elements in each period was equal to the number of electrons required to achieve the configuration of a noble gas ($1s^2$ for the first period and $ns^2np^6$ for all other periods, where $n$ is the period number). As shown in table 3, this number could be deduced using the Aufbau principle (see *2.1 Electron configuration*).

**Table 3.** The Aufbau principle and the periodic table

| Period number | 1 | 2 | 3 | 4 | 5 |
|---|---|---|---|---|---|
| Sublevels filled in period | 1s | 2s, 2p | 3s, 3p | 4s, 3d, 4p | 5s, 4d, 5p |
| Number of electrons in sublevels | 2 | 2 + 6 | 2 + 6 | 2 + 10 + 6 | 2 + 10 + 6 |
| Number of elements in period | 2 | 8 | 8 | 18 | 18 |

For example, the first energy level consists of a single 1s orbital that can hold up to two electrons. Therefore, the first period contains only two elements, hydrogen and helium. The second period contains eight elements, where the electrons fill one 2s and three 2p orbitals. The third period has the same length, as the electrons go to 3s and 3p orbitals. The five 3d orbitals are filled only in the fourth period, which contains ten more elements. These elements appear between the block of elements with outer s electrons and those with outer p electrons, giving the periodic table its characteristic shape. All groups are numbered from 1 to 18, which corresponds to the total number of elements in the fourth period (figure 9).

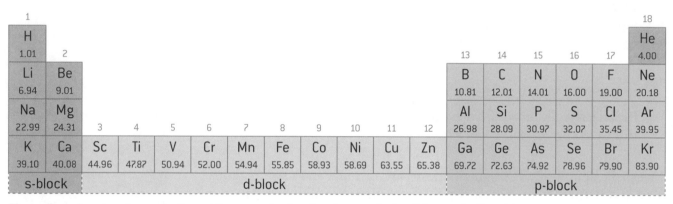

**Figure 9.** s, d and p elements form 18 groups in the first four periods of the periodic table

This arrangement of elements clearly shows the electron structure of their atoms: s orbitals are filled in groups 1 and 2, d orbitals in groups 3 to 12, and p orbitals in groups 13 to 18. For the same reason, the elements of groups 1 and 2 are called *s-block elements*, groups 3 to 12 contain *d-block elements*, and the last six groups contain *p-block elements*. The only exception is made for helium, which has outer s electrons but is placed in group 18 with other noble gases, all with full outer energy levels.

Energy sublevels 4f and 5f are filled in the sixth and seventh periods, respectively. Extending these periods by another fourteen elements would produce a very long table with 32 columns. Therefore, all f-elements are formally placed in the third group and shown in two separate rows below the main table (figure 10). A more detailed periodic table with element names is printed on the inside back cover of this book.

s-block

| 1 | | | | | | | | | | | | | | | | | 18 |
|---|---|---|---|---|---|---|---|---|---|---|---|---|---|---|---|---|---|
| 1<br>**H**<br>1.01 | 2 | | | | Key: | | | | | | p-block | | | | | | 2<br>**He**<br>4.00 |

Key:
atomic number
**Symbol**
relative atomic mass

| | | | | 13 | 14 | 15 | 16 | 17 | |
|---|---|---|---|---|---|---|---|---|---|

d-block

| Period | 1 | 2 | | | | | | | | | | | | 13 | 14 | 15 | 16 | 17 | 18 |
|---|---|---|---|---|---|---|---|---|---|---|---|---|---|---|---|---|---|---|---|
| 1 | 1<br>**H**<br>1.01 | 2 | | | | | | | | | | | | | | | | | 2<br>**He**<br>4.00 |
| 2 | 3<br>**Li**<br>6.94 | 4<br>**Be**<br>9.01 | | | | | | | | | | | | 5<br>**B**<br>10.81 | 6<br>**C**<br>12.01 | 7<br>**N**<br>14.01 | 8<br>**O**<br>16.00 | 9<br>**F**<br>19.00 | 10<br>**Ne**<br>20.18 |
| 3 | 11<br>**Na**<br>22.99 | 12<br>**Mg**<br>24.31 | 3 | 4 | 5 | 6 | 7 | 8 | 9 | 10 | 11 | 12 | | 13<br>**Al**<br>26.98 | 14<br>**Si**<br>28.09 | 15<br>**P**<br>30.97 | 16<br>**S**<br>32.07 | 17<br>**Cl**<br>35.45 | 18<br>**Ar**<br>39.95 |
| 4 | 19<br>**K**<br>39.10 | 20<br>**Ca**<br>40.08 | 21<br>**Sc**<br>44.96 | 22<br>**Ti**<br>47.87 | 23<br>**V**<br>50.94 | 24<br>**Cr**<br>52.00 | 25<br>**Mn**<br>54.94 | 26<br>**Fe**<br>55.85 | 27<br>**Co**<br>58.93 | 28<br>**Ni**<br>58.69 | 29<br>**Cu**<br>63.55 | 30<br>**Zn**<br>65.38 | 31<br>**Ga**<br>69.72 | 32<br>**Ge**<br>72.63 | 33<br>**As**<br>74.92 | 34<br>**Se**<br>78.96 | 35<br>**Br**<br>79.90 | 36<br>**Kr**<br>83.90 |
| 5 | 37<br>**Rb**<br>85.47 | 38<br>**Sr**<br>87.62 | 39<br>**Y**<br>88.91 | 40<br>**Zr**<br>91.22 | 41<br>**Nb**<br>92.91 | 42<br>**Mo**<br>95.96 | 43<br>**Tc**<br>(98) | 44<br>**Ru**<br>101.07 | 45<br>**Rh**<br>102.91 | 46<br>**Pd**<br>106.42 | 47<br>**Ag**<br>107.87 | 48<br>**Cd**<br>112.41 | 49<br>**In**<br>114.82 | 50<br>**Sn**<br>118.71 | 51<br>**Sb**<br>121.76 | 52<br>**Te**<br>127.60 | 53<br>**I**<br>126.90 | 54<br>**Xe**<br>131.29 |
| 6 | 55<br>**Cs**<br>132.91 | 56<br>**Ba**<br>137.33 | * | 72<br>**Hf**<br>178.49 | 73<br>**Ta**<br>180.95 | 74<br>**W**<br>183.84 | 75<br>**Re**<br>186.21 | 76<br>**Os**<br>190.23 | 77<br>**Ir**<br>192.22 | 78<br>**Pt**<br>195.08 | 79<br>**Au**<br>196.97 | 80<br>**Hg**<br>200.59 | 81<br>**Tl**<br>204.38 | 82<br>**Pb**<br>207.20 | 83<br>**Bi**<br>208.98 | 84<br>**Po**<br>(209) | 85<br>**At**<br>(210) | 86<br>**Rn**<br>(222) |
| 7 | 87<br>**Fr**<br>(223) | 88<br>**Ra**<br>(226) | ** | 104<br>**Rf**<br>(267) | 105<br>**Db**<br>(268) | 106<br>**Sg**<br>(269) | 107<br>**Bh**<br>(270) | 108<br>**Hs**<br>(269) | 109<br>**Mt**<br>(278) | 110<br>**Ds**<br>(281) | 111<br>**Rg**<br>(281) | 112<br>**Cn**<br>(285) | 113<br>**Nh**<br>(286) | 114<br>**Fl**<br>(289) | 115<br>**Mc**<br>(290) | 116<br>**Lv**<br>(293) | 117<br>**Ts**<br>(294) | 118<br>**Og**<br>(294) |

f-block

| * lanthanides | 57<br>**La**<br>138.91 | 58<br>**Ce**<br>140.12 | 59<br>**Pr**<br>140.91 | 60<br>**Nd**<br>144.24 | 61<br>**Pm**<br>(145) | 62<br>**Sm**<br>150.36 | 63<br>**Eu**<br>151.96 | 64<br>**Gd**<br>157.25 | 65<br>**Tb**<br>158.93 | 66<br>**Dy**<br>162.50 | 67<br>**Ho**<br>164.93 | 68<br>**Er**<br>167.26 | 69<br>**Tm**<br>168.93 | 70<br>**Yb**<br>173.05 | 71<br>**Lu**<br>174.97 |
|---|---|---|---|---|---|---|---|---|---|---|---|---|---|---|---|
| ** actinides | 89<br>**Ac**<br>(227) | 90<br>**Th**<br>232.04 | 91<br>**Pa**<br>231.04 | 92<br>**U**<br>238.03 | 93<br>**Np**<br>(237) | 94<br>**Pu**<br>(244) | 95<br>**Am**<br>(243) | 96<br>**Cm**<br>(247) | 97<br>**Bk**<br>(247) | 98<br>**Cf**<br>(251) | 99<br>**Es**<br>(252) | 100<br>**Fm**<br>(257) | 101<br>**Md**<br>(258) | 102<br>**No**<br>(259) | 103<br>**Lr**<br>(262) |

**Figure 10.** The IUPAC periodic table of the elements

## Question

9 Explore the first five periods of the periodic table in figure 10 and locate three pairs of elements where the order of relative atomic masses does not follow the order of atomic numbers. Suggest a reason for these exceptions.

In a modern periodic table, all elements are ordered by increasing atomic number and split into seven periods and eighteen groups. For historical reasons, the elements of the s- and p-blocks are often called *main-group elements*, in contrast to the *transition elements* of the d-block. The two series of f-block elements, *lanthanides* and *actinides*, are named after the first element in each series (lanthanum and actinium, respectively).

## Question

10 The series of lanthanides and actinides contain 15 elements each. Explain how the number of these elements relates to their electron structure.

**DP ready** | **Nature of science**

At present, the periodic table contains 118 elements, and this number is likely to grow in the future. However, the elements with atomic numbers over 100 decay so quickly — sometimes in milliseconds — that they have little practical use.

Several groups of elements have their own names. For example, group 1 elements except hydrogen are known as *alkali metals*, as they form very corrosive water-soluble hydroxides (alkalis) of the general formula MOH (M = Li, ..., Fr). Since the atoms of alkali metals have a single electron in their outer energy level, they form singly charged cations ($M^+$) in all their compounds.

The second group of the periodic table contains *alkaline earth metals*, which are named after their oxides (once called earths) of the general formula MO (M = Be, ..., Ra). The oxides and hydroxides of group 2 metals are only slightly soluble in water but still produce alkaline solutions. Because of their electron configuration ($ns^2$, where $n$ is the period number), alkaline earth metals tend to form doubly charged cations ($M^{2+}$) in chemical compounds.

Groups 3 to 12 contain transition elements, which are also known as *transition metals*. These groups are usually referred to by their numbers or by the name of their first member, such as "the scandium group", "the titanium group", and so on. Transition elements are less active than alkali and alkaline earth metals, and form a variety of ionic and covalent compounds, most of which are coloured.

Groups 13 to 15 are also named after their first members. With the exception of boron (B), the elements of group 13 are metals that tend to form triply charged ions ($M^{3+}$). Groups 14 and 15 contain both metals and nonmetals, the properties of which will be discussed later in this chapter.

Most members of group 16, sometimes referred to as *chalcogens*, are typical nonmetals. Since the atoms of these elements have electron configurations $ns^2np^4$, they need two more electrons to complete their outer energy levels. Therefore, they tend to accept electrons from other elements and form doubly charged anions, such as $O^{2-}$ and $S^{2-}$. At the same time, chalcogens can share their electrons and form covalent compounds.

Group 17 also consists of nonmetallic elements, known as *halogens*. Their outer energy levels contain seven electrons and have a configuration of $ns^2np^5$. As a result, all halogens are very reactive and readily form singly charged anions, such as $F^-$, $Cl^-$, $Br^-$ and $I^-$. In covalent compounds, halogens are usually monovalent.

Noble gases form the last group of the periodic table. As explained earlier, these elements have stable electron configurations ($1s^2$ for helium and $ns^2np^6$ for other noble gases), so they are unreactive and form very few chemical compounds.

**DP link**

In this book, transition elements will be discussed very briefly, but you will learn more about their properties in **13 The periodic table—the transition metals** in the IB Chemistry Diploma Programme.

**Internal link**

The bonding of chalcogen atoms was introduced in **1.2 Chemical substances, formulae and equations.**

**Question**

11 State the symbols and names for the following elements: **a)** the first d-block element of the fifth period; **b)** the lightest alkali metal; **c)** the second member of the carbon group; **d)** the last member of the lanthanide family; **e)** the halogen in the fourth period.

## Periodic properties of the elements

So far, we have used an intuitive approach to chemical properties of the elements. For instance, we have mentioned that some metals are "more active" than others, or that certain nonmetals "readily" form anions. These observations can be rationalized using certain quantitative characteristics of the elements, such as *atomic radius*, *ionization energy*, *electron affinity* and *electronegativity*.

 **Key term**

> The **atomic radius** (*r*) is the approximate distance from the atomic nucleus to the outer boundary of the electron cloud (figure 11).

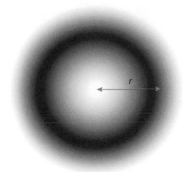

**Figure 11.** The atomic radius

Atomic radii of the elements demonstrate very clear periodic trends (figure 12). Moving to the right along each period, the electrons are added to the same energy level while the nuclear charge increases. As a result, the outer electrons are pulled closer to the nucleus, so the atomic radius decreases across periods. In contrast, when we move down the group, the electrons are added to higher and higher energy levels and thus get further away from the nucleus. As a result, the atomic radius increases down groups.

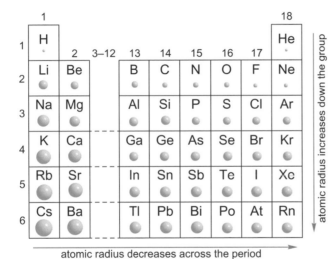

**Figure 12.** Relative atomic radii of the elements

 **Key term**

> The **ionization energy** (*E*$_i$) of an element is the energy required to remove the outermost electron from an atom of that element in the gaseous state:
>
> $$X(g) + E_i \rightarrow X^+(g) + e^-$$

Since the ionization energy of a single atom has a very small value, it is usually multiplied by the Avogadro constant (*1.3 Stoichiometric relationships*) and expressed in kJ mol$^{-1}$. All elements have positive $E_i$ values, which means that the energy is always consumed when a neutral atom loses an electron and forms a cation.

Ionization energies are inversely related to atomic radii, as larger atoms tend to have lower $E_i$ values. Indeed, the further away the electrons are from the atomic nucleus, the easier it is to pull them off the atom.

**Internal link**

You can review a discussion of the Avogadro constant in **1.3 Stoichiometric relationships**.

Therefore, the ionization energy generally increases across periods and decreases down groups (figure 13).

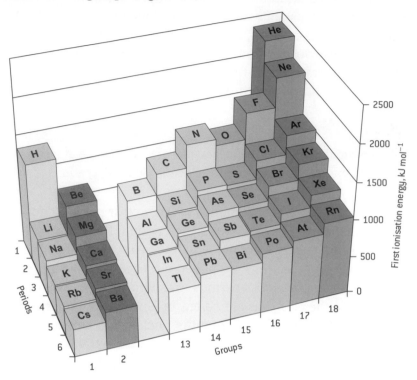

**Figure 13.** First ionization energies of the elements

The ionization energy of an element can provide some information on its chemical activity. Alkali metals (group 1) and alkaline earth metals (group 2) have very low ionization energies, so they readily lose electrons and form cations. Caesium (Cs) has the lowest ionization energy (376 kJ mol⁻¹) and is so reactive that it spontaneously ignites in air and explodes on contact with water. On the other hand, metals with high ionization energies are usually unreactive and tend to form covalent compounds rather than cations. One of the most chemically resistant metals, platinum (Pt), has an $E_i$ value of 864 kJ mol⁻¹, more than double that for caesium.

Nonmetals have very high ionization energies, often above 1000 kJ mol⁻¹, so they never form cations in chemical reactions. The $E_i$ values of helium (2372 kJ mol⁻¹) and neon (2081 kJ mol⁻¹) are so high that these noble gases are completely unreactive. Heavier noble gases (krypton, xenon and radon) have somewhat lower ionization energies, so they can form chemical compounds with fluorine and some other nonmetals.

 **Key term**

The ability of an element to form anions is characterized by its **electron affinity ($E_{ea}$)**, which is the energy released when an electron is added to a neutral atom of that element in the gaseous state:

$$X(g) + e^- \rightarrow X^-(g) + E_{ea}$$

In contrast to ionization energy, which is always positive, the electron affinity of an element can be either negative or positive. A negative $E_{ea}$ value means that the energy is released when the atom accepts an electron. Halogens (group 17) and chalcogens (group 16) have highly negative electron affinities (between –350 and –140 kJ mol⁻¹), so they are very reactive and readily form anions. In contrast, most metals have either positive or slightly negative electron affinities, so their anions are unstable. Finally, the $E_{ea}$ values for noble gases cannot be determined, as their atoms have complete outer energy levels and thus cannot accept any more electrons.

The periodic trends in electron affinity are less clear than those for ionization energy and atomic radius. Overall, the $E_{ea}$ values increase across periods but show no common pattern in groups (figure 14). Please note that each bar in figure 14 represents a negative value, so the higher the bar, the lower (more negative) the electron affinity is.

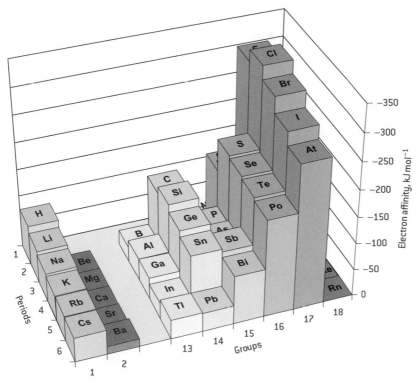

**Figure 14.** Electron affinities of the elements (bars represent negative values)

The ionization energy and electron affinity characterize isolated atoms of chemical elements and their ability to form ions. The behaviour of elements in covalent molecules is better described by the concept of *relative electronegativity* ($\chi$), introduced by the American scientist Linus Pauling in 1932.

 **Internal link**

Covalent bonding is introduced in **1.2 Chemical substances, formulae and equations**.

 **Key term**

Relative **electronegativity** ($\chi$) is the ability of an element to attract electrons in a covalent bond.

A covalent bond is formed by a pair of electrons that belongs to two (or sometimes more) atoms at the same time. If both atoms are identical, the electron cloud of shared electrons has the highest density at the same distance from each atom. Such a bond is called a *non-polar covalent bond*, as neither atom develops a net electric charge. For example, the covalent bonds in both hydrogen, $H_2$, and chlorine, $Cl_2$, are non-polar, so the shared electron pairs in these molecules are shown as two dots exactly halfway between the atoms:

$$H\cdot \ + \ \cdot H \rightarrow H:H$$
$$Cl\cdot \ + \ \cdot Cl \rightarrow Cl:Cl$$

In contrast, the atoms of different elements tend to form *polar covalent bonds*, where the shared electron pair is shifted towards the more electronegative element. For example, the formation of hydrogen chloride, HCl, can be represented as follows:

$$H\cdot \ + \ \cdot Cl \rightarrow H:Cl$$

The shared electron pair in hydrogen chloride is pulled towards chlorine, showing that chlorine is more electronegative than hydrogen.

As the electron density around the chlorine atom increases, chlorine develops a partial negative charge ($\delta-$). At the same time, the electron density around the hydrogen atom decreases, so hydrogen develops a partial positive charge ($\delta+$).

By measuring the values of $\delta+$ and $\delta-$ charges in hydrogen chloride and other molecules, it is possible to calculate relative electronegativities ($\chi$) of all elements that can form covalent bonds (figure 15). For example, the $\chi$ values for hydrogen and chlorine are 2.2 and 3.2, respectively. As suggested by the name, relative electronegativities are unitless. However, the word "relative" is often omitted.

## Maths skills: Partial and full charges

It is important to understand the meaning of the lowercase Greek letter $\delta$ (delta) in $\delta+$ and $\delta-$. This symbol means "partial". Partial charges occur on atoms in covalent molecules and have non-integer values, typically much less than 1. For example, the partial charges of hydrogen and chlorine in the HCl molecule are 0.15+ and 0.15−, respectively.

In contrast, the charges on ions in ionic compounds are assumed to have integer values, such as 1−, 2+ and so on, so they are shown without the $\delta$ symbol.

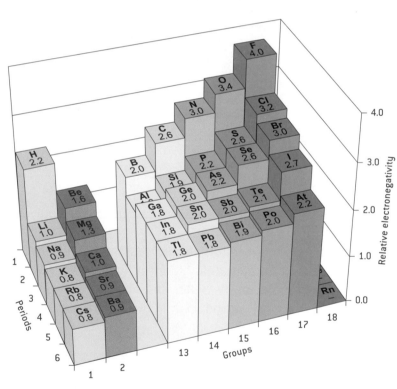

**Figure 15.** Relative electronegativities ($\chi$) of the elements

As shown in figure 15, relative electronegativities demonstrate clear periodic trends, generally increasing across periods and decreasing down groups. The most electronegative element, fluorine, occupies the top right corner of the periodic table and has a $\chi$ value of 4.0. The least electronegative element with $\chi$ = 0.8, caesium, is located in the opposite corner. Noble gases are usually not assigned any $\chi$ values.

### Question

12 Using partial charges, show the distribution of electron density between hydrogen and oxygen in a molecule of water.

## Oxidation numbers

The concept of electronegativity can be used to describe covalent molecules as if they were ionic compounds. To do this, each atom in a covalent molecule is assigned an *oxidation number*, which is numerically equal to the ionic charge this atom would have in a similar molecule where all polar covalent bonds are formally replaced with ionic bonds. For example, if we imagine that the shared electron pair in hydrogen chloride is completely transferred to chlorine, the atom of hydrogen will become a cation, $H^+$, with an ionic charge of $1+$, and the atom of chlorine will become an anion, $Cl^-$, with the ionic charge of $1-$. Therefore, the oxidation numbers of hydrogen and chlorine in HCl are $+1$ and $-1$, respectively.

To distinguish between ionic charges and oxidation numbers, the signs in oxidation numbers are written before the digits, and the digit "1" in oxidation numbers $+1$ and $-1$ is never omitted. In truly ionic compounds, such as sodium chloride, the oxidation numbers of ions are equal to their actual charges.

In all cases, the more electronegative element gets a negative oxidation number, and the less electronegative element gets a positive oxidation number.

Finally, the oxidation numbers of all atoms in elementary substances are assumed to be 0 (zero).

These rules are illustrated by the examples in table 4.

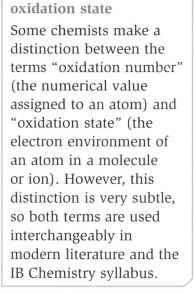

**DP ready** — **Nature of science**

**Oxidation number and oxidation state**

Some chemists make a distinction between the terms "oxidation number" (the numerical value assigned to an atom) and "oxidation state" (the electron environment of an atom in a molecule or ion). However, this distinction is very subtle, so both terms are used interchangeably in modern literature and the IB Chemistry syllabus.

**Table 4.** Covalent and ionic bonding, oxidation numbers and charge

| Compound | Bonding type | Ionic charges | Oxidation numbers |
|---|---|---|---|
| $Cl_2$ | non-polar covalent | — | 0 |
| HCl | polar covalent | — | $+1$ (H) and $-1$ (Cl) |
| NaCl | ionic | $Na^+$ and $Cl^-$ | $+1$ (Na) and $-1$ (Cl) |
| $H_2S$ | polar covalent | — | $+1$ (H) and $-2$ (S) |
| $Na_2S$ | ionic | $Na^+$ and $S^{2-}$ | $+1$ (Na) and $-2$ (S) |

If two elements in a compound have identical relative electronegativities (for example, hydrogen and phosphorus in phosphine, $PH_3$), their oxidation states become ambiguous. Most chemists assume that hydrogen has an oxidation state of $+1$ in all compounds with nonmetals and metalloids, so the oxidation state of phosphorus in phosphine is $-3$. Note that in compounds with metals, hydrogen and other nonmetals always have negative oxidation states.

### Worked example: Determining oxidation numbers

**1.** Determine the oxidation number of both elements in water, $H_2O$.

*Solution*

The less electronegative element gets a positive oxidation number, so hydrogen will have an oxidation number of $+1$. As there are two hydrogen atoms to one of oxygen, the latter has an oxidation number of $-2$.

13 Determine the oxidation numbers of all elements in the following substances:
a) $MgCl_2$; b) $NH_3$; c) $N_2$; d) $KH$; e) $CO_2$.

Typical oxidation states of most elements can be deduced from their positions in the periodic table. In groups 1–7, the highest oxidation state of an element is equal to its group number, and in groups 12–17 it can be found as the group number minus 10. For example, sulfur is a group 16 element, so its highest oxidation state is $16 - 10 = +6$. In groups 8–11, the oxidation states of elements are less predictable and can vary from +2 to +8. Noble gases (group 18) are unreactive, although xenon forms several compounds with oxidation numbers of up to +8.

The most electronegative element, fluorine, cannot have a positive oxidation number, as no other element has a greater attraction to a shared electron pair. Therefore, fluorine is assigned an oxidation number of –1 in all its compounds. Similarly, the second most electronegative element, oxygen, has positive oxidation numbers only in compounds with fluorine, such as $OF_2$ and $O_2F_2$. In almost all other compounds, oxygen has an oxidation number of –2.

Negative oxidation states are typical for nonmetals of groups 14 to 17. The lowest (most negative) oxidation number of an element can be found as its group number minus 18. For example, fluorine is a group 17 element, so its lowest oxidation number is $17 - 18 = -1$. The only nonmetal of group 1, hydrogen, can also have an oxidation number of –1. Metals and noble gases never have negative oxidation numbers in their compounds.

14 Deduce the highest and lowest oxidation states for the following elements: a) magnesium; b) chromium; c) phosphorus; d) iodine.

## Metals, nonmetals and metalloids

The elements in the periodic table are traditionally divided into metals, nonmetals and metalloids. Although we have used some of these terms already, it is now time to define them in a more systematic way.

*Metals*, or *metallic elements*, are characterized by low values of electronegativity (typically below 2.0), low positive ionization energy and positive or slightly negative electron affinity. As a result, metals readily lose electrons to more electronegative elements and form cations. In covalent compounds, metals have positive oxidation states. Metals do not react with one another, although they often form homogeneous mixtures (alloys).

In contrast, *nonmetals* have high values of electronegativity (2.2 or above), high ionization energy and large negative electron affinity. In reactions with metals, they tend to accept electrons and form anions. Most nonmetals can also react with one another, producing covalent compounds. Although noble gases are formally classified as nonmetals, they are very unreactive and thus lack both metallic and nonmetallic properties.

*Metalloids* are a group of seven chemical elements that show a mixture of metallic and nonmetallic properties. These elements have intermediate values of electronegativity (2.0–2.2), ionization energy and electron affinity. Metalloids do not form ions, have negative oxidation states in compounds with metals and positive oxidation states in compounds with nonmetals. In the periodic table, metalloids form a diagonal line that separates metals from nonmetals (figure 16).

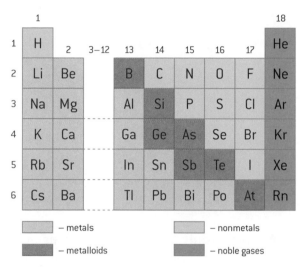

**Figure 16.** Classification of chemical elements

As can be seen from figure 16, metallic properties of the elements decrease across periods and increase down groups, which reflects the trend in electronegativity. Typical nonmetals (other than hydrogen) are located in the top right corner of the periodic table, with fluorine being both the most electronegative element and the most active nonmetal. Almost all other elements in the periodic table are metals, which are separated from nonmetals by a narrow band of metalloids.

> **Question** ⓠ
>
> 15 Element 84, polonium, Po, is classified in some textbooks as the eighth metalloid. Consider the properties, electron structure and position of this element in the periodic table and give at least three reasons for the classification.

We will continue the discussion of metals and nonmetals in the next topic, where you will see how the nature of chemical bonding can be explained in terms of electron structure and electronegativity of the elements.

## 2.3 Chemical bonding

### Bond types

Chemical bonds are formed when atoms share their electrons or transfer them to one another. In metals and covalent molecules, positively charged atomic nuclei are attracted to negatively charged clouds of shared electrons. In ionic compounds, the oppositely charged ions are attracted to one another. Therefore, all chemical bonds are electromagnetic in nature, as they involve interactions between charged species.

 **Internal link**

Chemical bonds are traditionally classified as covalent, ionic and metallic: see **1.2 Chemical substances, formulae and equations**.

Covalent bonds can be polar or non-polar (see *2.2 The periodic law*). Common types of chemical bonding are shown in figure 17.

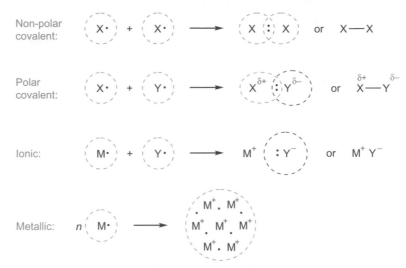

**Figure 17.** Common types of chemical bonding (X, Y = nonmetals, M = metal)

In elementary substances, only two types of bonding are possible: metallic (in metals) and non-polar covalent (in nonmetals). The bonding in metalloids is intermediate between metallic and covalent. Noble gases are unreactive, so their atoms do not form any chemical bonds with one another.

The type of chemical bonding between two different elements depends on their relative electronegativities. If these electronegativities differ by more than 1.8 units, the valence electrons are completely transferred to the more electronegative element, producing an ionic bond. This situation is typical for compounds formed by an active metal and an active nonmetal. For example, the difference in electronegativity between sodium ($\chi$ = 0.9) and chlorine ($\chi$ = 3.2) is 3.2 – 0.9 = 2.3, so sodium chloride is an ionic compound.

A smaller difference in electronegativity, from 0.3 to 1.7 units, is typical for polar covalent bonds. Such bonds are often formed between two nonmetals. For example, in hydrogen chloride $\chi$(H) = 2.2 and $\chi$(Cl) = 3.2, so $\Delta\chi$ = 3.2 – 2.2 = 1.0. Therefore, the bond between hydrogen and chlorine is polar covalent.

Elements with similar electronegativities ($\Delta\chi$ < 0.3) form either non-polar covalent or metallic bonds. Obviously, the bond between two metals will be metallic while the bond between two nonmetals with approximately the same $\chi$ values will be non-polar covalent.

**Question**

16 Using Figure 15, determine the bonding types in the following compounds and materials: **a)** magnesium chloride; **b)** nitrogen tribromide; **c)** alloy of aluminium and beryllium; **d)** carbon dioxide.

When discussing chemical bonding, it is important to understand that "pure" ionic and metallic bonds do not exist in nature. All such bonds have partial covalent character, as the electron density between interacting atoms or ions never falls to zero. Even the elements with the greatest difference in electronegativity, caesium and fluorine, form bonds that are approximately 92% ionic and 8% covalent. Similarly, metals and their alloys show varying degrees of metallic and covalent bonding. Nevertheless, the concepts of ionic and metallic bonding are useful theoretical models that provide a simple classification of real substances and correctly describe their physical and chemical properties.

The strength of a chemical bond can be defined as the energy required for breaking one mole of covalent bonds or pulling apart one mole of ions or metal atoms. In most cases, this energy varies from 200 to 800 kJ mol$^{-1}$ for covalent and metallic substances and from 600 to 3500 kJ mol$^{-1}$ for ionic compounds. These values are relatively large, so covalent, metallic and ionic bonding are collectively known as *strong*, or *primary*, *bonds*.

### Intermolecular forces

In addition to this strong bonding, atoms, ions and molecules can also participate in weaker interactions, known as *secondary bonds*. These interactions are often called *intermolecular forces*, as they are primarily responsible for holding together the molecules or other structural units of chemical substances. The most common types of secondary bonding are hydrogen bonds and van der Waals forces.

A *hydrogen bond* typically involves electrostatic attraction between two molecules or ions, one of which has a hydrogen atom with a partial positive charge while another has a partially negatively charged atom (figure 18). To have a significant positive charge, the hydrogen atom must be covalently bonded to a highly electronegative element, such as fluorine, oxygen or nitrogen. The negatively charged atom of the second molecule must also be fluorine, oxygen or nitrogen, as only these three elements have sufficient electronegativity to attract the electron density from other atoms. Therefore, hydrogen bonds are typical for hydrogen fluoride, water, ammonia and other molecules with O–H and N–H groups.

**Figure 18.** Hydrogen bonds (dotted lines) in hydrogen fluoride (left) and water (right)

 **Internal link**

Bond strengths (bond enthalpies) are defined and discussed in **5.1 Thermochemistry**.

 **Key term**

**Intermolecular forces** are the bonds between (rather than within) molecules or other structural units. There are two kinds of intermolecular forces, hydrogen bonding and van der Waals forces, the latter including London dispersion forces.

 **Key term**

A **hydrogen bond** is a secondary bond formed between a hydrogen atom with a partial positive charge on one species and a negatively charged atom of F, N or O on another species.

### Question

17 Draw hydrogen bonds between the following molecules and ions: **a)** $NH_3$ and $NH_3$; **b)** HF and $H_2O$; **c)** HF and F$^-$.

Although hydrogen bonds are relatively weak (typically 5–40 kJ mol⁻¹), they have a large effect on the physical properties of compounds. In particular, the boiling points of hydrogen fluoride and water are significantly higher than those of other hydrides of groups 16 and 17 elements (figure 19). Similar trends are observed for melting points and aqueous solubility of these compounds.

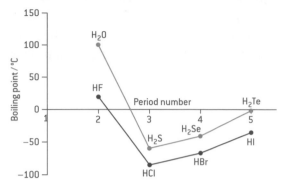

**Figure 19.** Boiling points of hydrides of group 16 and 17 elements

## DP link

Hydrogen bonds play an important role in supporting the three-dimensional structures of proteins, DNA and other biological molecules, which are discussed in **B Biochemistry** in the IB Chemistry Diploma Programme.

## Key term

**Van der Waals forces** are weak forces between molecules caused by uneven distribution of the electric charge within the molecules.

An **electric dipole** is the separation of negative and positive charges within a molecule. Molecules with electric dipoles are called **polar**.

## Question

18 Suggest, with a reason, which of the two substances $NH_3$ and $PH_3$ has a higher boiling point.

Molecules that cannot form hydrogen bonds are held together only by very weak *van der Waals forces*, which are named after the Dutch physicist Johannes van der Waals. These forces are caused by uneven distribution of electric charge in molecules, either due to the difference in electronegativity between the elements or random movement of electron clouds. For example, the molecule of hydrogen chloride can be represented as an *electric dipole* with a positive charge on the hydrogen atom and a negative charge on the chlorine atom:

Such molecules are called *polar*, in contrast to *non-polar* molecules with symmetrical distribution of positive and negative charges. When two polar molecules come close to each other, the oppositely charged ends of their dipoles experience electrostatic attraction, so the molecules change their orientation to maximize the attraction:

This kind of electrostatic attraction, known as *dipole–dipole interaction*, occurs only in molecules with permanent dipoles, such as hydrogen chloride. In solutions of ionic compounds, polar molecules can also participate in *ion–dipole interactions* with cations and anions:

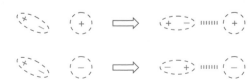

Non-polar molecules can form temporary, or *induced*, *dipoles* when their electron clouds are disturbed by other molecules or ions. For example, the positive end of an approaching polar molecule can pull the electron density of a non-polar molecule to one side, creating an induced dipole:

Similarly, the negative end of an approaching polar molecule will push the electron density of a non-polar molecule to the opposite side, creating another induced dipole:

In both cases, the orientation of the induced dipole will match the orientation of the approaching permanent dipole, so both molecules will be attracted to each other by electrostatic force. The resulting *dipole–induced dipole* interaction will be weaker than the interaction between two permanent dipoles, as the displacement of electrons in the second molecule will require some energy.

The weakest of all van der Waals forces are the interactions between non-polar molecules, known as *London dispersion forces*. Named after the German physicist Fritz London, these forces arise from random movement of electrons in molecules or individual atoms. Although the average distribution of electron density in a non-polar molecule is symmetrical, at any given time the electron cloud can shift slightly to one side of the molecule, creating a short-lived *instantaneous dipole*:

Once formed, the instantaneous dipole will disturb the electron clouds of neighbouring molecules in the same way as any other dipole, creating induced dipoles:

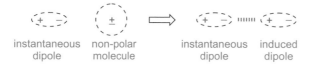

The resulting *instantaneous dipole–induced dipole* interactions are so weak that they can normally be detected only in non-polar molecules. However, London dispersion forces exist in all substances, regardless of their polarity or bonding type. The strength of London dispersion forces increases with the number of electrons and the overall size of the molecule, so these forces can become very significant in large molecules or even individual atoms of heavy elements.

For example, the boiling points of monoatomic noble gases increase down the group, along with the number of electrons in their atoms (figure 20). Similar trends are observed for halogens and other substances composed of non-polar molecules.

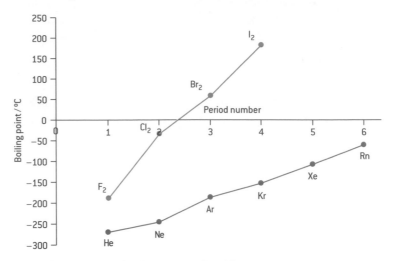

**Figure 20.** Boiling points of halogens and noble gases

**Internal link**

Ion–dipole interactions will be discussed in **4.3 Solutions of electrolytes**.

Common types and strengths of intermolecular forces are summarized in table 5. Ion–dipole interactions are not included in this table, as their strength varies greatly among compounds.

**Table 5.** Typical strengths of intermolecular forces

| Interaction type | | Strength / kJ mol$^{-1}$ |
|---|---|---|
| hydrogen bonding | | 5–40 |
| van der Waals forces | dipole–dipole | 3–25 |
| | dipole–induced dipole | 1–20 |
| | London dispersion | 1–10 |

If more than one type of intermolecular force is present in the same substance, we usually need to consider only the strongest interaction. For example, water forms hydrogen bonds, so it has a higher boiling point than hydrogen sulfide (figure 19), despite the fact that molecules of hydrogen sulfide have more electrons and thus experience stronger London attraction.

**Question**

19 Identify the intermolecular forces in the following substances or mixtures: **a)** argon; **b)** hydrogen fluoride; **c)** hydrogen chloride; **d)** a mixture of argon and hydrogen chloride.

We have now reviewed all common types of chemical bonding and shown how its nature can be predicted from the electron configurations and electronegativities of participating elements. In the final sections of this chapter, we will discuss the effects of chemical bonding on the structure and properties of ionic, covalent and metallic substances.

## Ionic compounds

Cations and anions in solid ionic compounds form highly symmetrical three-dimensional structures, known as *crystal lattices* (figure 21). Each cation in a crystal lattice is surrounded by anions, and each anion is surrounded by cations. This arrangement increases the electrostatic attraction between oppositely charged ions and decreases the repulsion between ions of the same charge.

It is important to understand that electrostatic forces are non-directional, so each ion interacts not only with its neighbours but also with all other ions in the crystal. To emphasize this fact, ionic compounds are often said to have *giant ionic structures* or *giant ionic lattices*. For the same reason, ionic bonds should never be shown as lines in structural formulae. For example, sodium chloride can be represented as $Na^+Cl^-$ but **not** as "Na—Cl".

Since the ions are held together by strong electrostatic forces, most ionic compounds have high melting and boiling points. For instance, sodium chloride melts at 801°C and boils at 1413°C. According to Coulomb's law, the strength of electrostatic attraction is proportional to the charges of cations and anions and inversely proportional to the distance between these ions. As a result, melting and boiling points of ionic compounds increase with ionic charge and decrease with ionic size. For example, magnesium oxide contains doubly charged ions ($Mg^{2+}$ and $O^{2-}$), so its melting and boiling points (2825 and 3600°C, respectively) are much higher than those of sodium chloride, which is composed of singly charged ions ($Na^+$ and $Cl^-$). Similarly, melting and boiling points of sodium halides decrease as the size of the halide ion increases from $F^-$ to $I^-$ (figure 22).

**Internal link**

Ionic lattices are introduced in **1.2 Chemical substances, formulae and equations**.

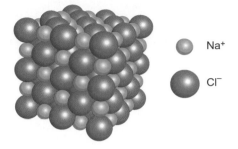

**Figure 21.** Crystal lattice of sodium chloride

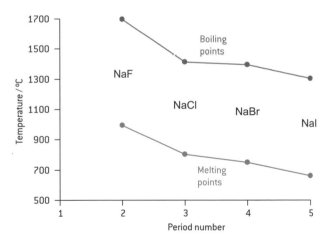

**Figure 22.** Melting and boiling points of sodium halides

## Question

**20** Compare the melting points of the following ionic compounds:
**a)** NaCl and KCl; **b)** $CaCl_2$ and CaO; **c)** MgO and CaO.

In the solid state, all ions have fixed positions, so they cannot move without breaking the lattice. As a result, solid ionic compounds are hard and brittle and do not conduct electricity. In contrast, molten ionic compounds are good conductors, as their ions are mobile and can carry electric charge.

Most ionic compounds are soluble in water and other polar solvents, but do not dissolve in non-polar solvents, such as hydrocarbons.

**Internal link**

The behaviour of ionic compounds in aqueous solutions will be discussed in **4 Solutions and concentration**.

## Covalent substances and Lewis structures

As you already know, covalent bonds are formed when atoms share their electrons. In contrast to ionic bonding, covalent bonds are directional, so in structural formulae they are usually drawn as lines connecting specific atoms. Each line corresponds to one shared electron pair, which can also be shown as two dots between the atoms. For example, the structure of hydrogen fluoride can be represented as follows:

$$\text{H—F} \quad \text{or} \quad \text{H : F}$$

So far, we have used lines or dots only for shared, or *bonding* electron pairs. However, the atoms of most chemical elements may also have *lone*, or *non-bonding*, electron pairs, which can also be shown in structural formulae. For example, fluorine is a group 17 element, so it has seven electrons in its outer energy level. By sharing one electron with hydrogen, fluorine achieves a stable eight-electron configuration (an octet):

According to the orbital diagram, these eight electrons form four electron pairs. One of these pairs is bonding, as it is shared between fluorine and hydrogen, while the other three pairs are lone (non-bonding), as they belong to fluorine only. At the same time, hydrogen has no lone pairs, as its outer energy level can hold only the two electrons provided by the shared pair. Therefore, we can rewrite the structural formula of hydrogen fluoride as follows:

$$\text{H} \!—\! \overset{\displaystyle ..}{\underset{\displaystyle ..}{\text{F}}}\text{:} \qquad \text{or} \qquad \text{H} \text{:} \overset{\displaystyle ..}{\underset{\displaystyle ..}{\text{F}}}\text{:}$$

Such formulae, known as *Lewis structures*, were proposed in 1916 by the American physical chemist Gilbert Lewis. As you will see shortly, Lewis structures can be used for predicting the three-dimensional shapes of covalent molecules.

**Figure 23.** Gilbert Lewis (1875–1946)

### Question

**21** Draw the Lewis structures for the molecules of hydrogen ($H_2$) and fluorine ($F_2$).

When drawing Lewis structures for larger molecules, use the following strategy.

1. Determine the total number of valence electrons in the molecule. For example, a molecule of water contains two hydrogen atoms (one valence electron each) and one oxygen atom (six valence electrons), so the total number of valence electrons is $1 \times 2 + 6 = 8$:

$$\text{H}\cdot \qquad \cdot\overset{\displaystyle ..}{\underset{\displaystyle ..}{\text{O}}}\cdot \qquad \cdot\text{H}$$

2. Connect all atoms by single bonds and subtract two valence electrons for each bond from their total number. In a water molecule, there are two O–H bonds, so $8 - 2 \times 2 = 4$ electrons will remain non-bonding:

$$\text{H—O—H}$$

3. Arrange the remaining electrons so that each atom has a complete outer level. These electrons may either stay on individual atoms as lone pairs or be shared between atoms and form additional bonds. If more than one combination of electrons is possible, add lone pairs to more electronegative atoms first. In our example, the four remaining electrons can be placed only on the oxygen atom, as the valence levels of both hydrogen atoms are already full:

$$H - \overset{..}{\underset{..}{O}} - H$$

4. Check again whether all non-hydrogen atoms satisfy the octet rule, and each hydrogen atom forms a single bond. In our case, the oxygen atom has eight electrons (four shared and four non-bonding), and both hydrogen atoms are bonded as required, so the Lewis structure is correct.

**Question**

22 Draw the Lewis structures for ammonia ($NH_3$) and methane ($CH_4$).

**Worked example: Drawing Lewis structures**

**2.** Draw the Lewis structure for the molecule of oxygen ($O_2$).

*Solution*

The $O_2$ molecule has twelve valence electrons. A single bond between the oxygen atoms will leave ten. If we draw these ten electrons as five lone pairs, one of the oxygen atoms will not satisfy the octet rule, as it will have six electrons instead of eight:

$$:\overset{..}{\underset{..}{O}} - \overset{.}{\underset{.}{O}}:$$

Therefore, we need to rearrange the electrons by sharing one more pair between the oxygen atoms giving a double bond:

$$:\overset{.}{\underset{.}{O}} = \overset{.}{\underset{.}{O}}:$$

Now each oxygen atom has eight electrons (four shared and four non-bonding), so the Lewis structure is correct.

**Question**

23 Draw the Lewis structures for nitrogen ($N_2$) and carbon dioxide ($CO_2$).

Several elements of groups 2 and 13, such as beryllium, boron and aluminium, do not always satisfy the octet rule and may have incomplete valence levels in covalent molecules. For example, beryllium dihydride and boron trifluoride have the following Lewis structures:

$$H - Be - H$$

In these molecules, beryllium has only four electrons in the valence level (two shared pairs) and boron has six electrons (three shared pairs). Such *electron-deficient* species are very reactive and tend to form additional covalent bonds with other molecules or ions. In particular, boron trifluoride can react with a fluoride anion, $F^-$, producing a tetrafluoroborate anion, $BF_4^-$:

$$BF_3 + F^- \rightarrow BF_4^-$$

The fluoride anion has a full octet of electrons, arranged as four lone pairs. At the same time, the boron atom in $BF_3$ has only six electrons, so one of its atomic orbitals is empty. Therefore, the fluoride anion can share one of its lone electron pairs with boron, producing a covalent bond (figure 24):

Figure 24. Formation of a coordinate bond in the boron tetrafluoride anion

In this reaction, fluorine acts as a donor of an electron pair while boron accepts this pair to an empty atomic orbital and completes its outer electron level. Covalent bonds formed in this way are called *coordinate bonds*. Note that coordinate bonds differ from "normal" covalent bonds only by the mechanism of their formation but not by their properties: all four B–F bonds in $BF_4^-$ have identical lengths and thus are indistinguishable from one another.

> **Question**
>
> 24 Draw the Lewis structures for the following species: **a)** $BeCl_2$; **b)** $BH_3$; **c)** $BH_4^-$. Identify the atoms that do not obey the octet rule.

There are many exceptions to the octet rule beside beryllium and group 13 elements. For example, elements of the third and further periods can accommodate more than eight electrons in their valence level by using vacant 3d orbitals.

### VSEPR theory

Lewis structures are drawn in two dimensions (on paper), so they tell us very little about three-dimensional shapes of molecules and ions. However, these shapes can be deduced from Lewis structures using the *valence-shell electron-pair repulsion (VSEPR) theory*. According to this theory, covalent bonds and lone electron pairs are treated as separate *electron domains*, which are regions of space with increased electron density. Double and triple bonds are counted as single electron domains because all shared electrons of such bonds are located in the same region of space.

Because of their negative charges, electron domains repel one another, so they spread out as far as possible to minimize this repulsion. In a sense, electron domains behave like inflated balloons tied together, where the tie point represents the atomic nucleus (figure 25).

Figure 25. Balloons as models of electron domains

## Key term

A **coordinate bond**, sometimes called a dative bond, is a covalent bond in which both electrons in the shared pair are provided by the same atom. Once formed they are no different from covalent bonds formed by sharing one electron from each atom.

## DP link

In some molecules, such as NO and $NO_2$, the total number of valence electrons is odd, so at least one electron remains unpaired. The Lewis structures of these and many other species will be discussed in **14 Chemical bonding and structure** in the IB Chemistry AHL Diploma Programme.

## Key term

**Valence-shell electron-pair repulsion (VSEPR) theory** treats covalent bonds and lone pairs as separate electron domains that spread out around the nucleus so as to minimize repulsion. VSEPR theory enables us to predict and explain the shapes of molecules.

Two balloons tied together will point in opposite directions, forming a straight line (angle 180°). Three balloons will point to the corners of a planar triangle, positioned at 120° from one another. Four balloons will point to the corners of a tetrahedron with angles of 109.5° about the central tie. Similarly, atoms with two, three and four electron domains will adopt a linear, trigonal planar and tetrahedral configurations, respectively (table 6).

**Table 6.** Molecular geometries of atoms with two, three and four electron pairs

| Number of electron domains | Molecular geometry | Molecular structure | Bond angle | Example |
|---|---|---|---|---|
| 2 | linear | | 180° | $BeH_2$ |
| 3 | trigonal planar | | 120° | $BH_3$ |
| 4 | tetrahedral | | 109.5° | $CH_4$ |

The bond angles given in table 6 are valid only for symmetrical molecules with no lone electron pairs on the central atom. Lone pairs occupy more space and cause greater repulsion than bonding pairs, so they reduce the angles between the bonding pairs by several degrees. For example the H–N–H bond angle in ammonia is 107.8° instead of 109.5° because of the presence of a lone electron pair on the nitrogen atom. Two lone electron pairs on the oxygen atom in water reduce the H–O–H bond angle even further to 104.5° (figure 26). You do not need to memorize these values — it is sufficient to remember that lone pairs have small but noticeable effects on bond angles.

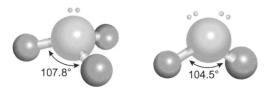

**Figure 26.** Bond angles in ammonia (left) and water (right)

## Question

25 Deduce the shapes of the following species and describe them in terms of electron domains and molecular geometry: a) $BeCl_2$; b) $NF_3$; c) $BF_3$; d) $CO_2$; e) $BF_4^-$.

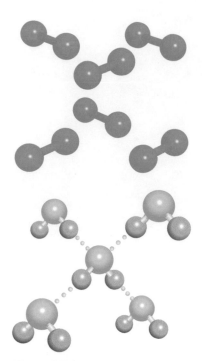

**Figure 27.** Structures of molecular solids: iodine (top) and water ice (bottom); hydrogen bonds in ice are shown as blue dots

## Properties of covalent substances

In the solid and liquid states, small covalent molecules are held together by weak van der Waals forces and, in some cases, by hydrogen bonds (figure 27). Since these forces are relatively weak, covalent substances with simple molecular structures tend to have low melting and boiling points, low density and high volatility. Under normal conditions, many of these substances are gases (such as oxygen, ammonia and carbon dioxide) or liquids (such as bromine and water). Molecular solids, such as iodine or white phosphorus, are soft, volatile and do not conduct electricity.

In other substances, known as *covalent network solids*, all atoms are joined together by a continuous network of covalent bonds. Such substances are said to have *giant covalent structures*, in contrast to the *simple molecular structures* of iodine or water.

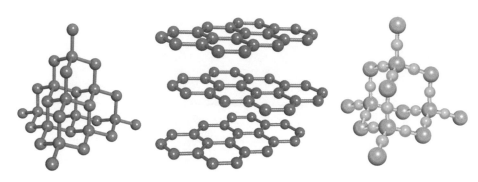

**Figure 28.** Structures of covalent network solids: diamond (left), graphite (middle) and silicon dioxide (right)

Almost all covalent network substances have high melting and boiling points, and are hard, brittle, non-volatile under normal conditions, and do not conduct electricity in either the solid or liquid state. Diamond, an elementary form of carbon, is one of the hardest known materials, as any deformation of its lattice can only be achieved by breaking the very strong covalent bonds between carbon atoms. All bond angles in diamond are 109.5°, so each carbon atom has a tetrahedral configuration (figure 28). Another material with a diamond-like structure is silicon dioxide, $SiO_2$, where silicon and oxygen atoms form a three-dimensional network of covalent bonds. Similar to diamond, the crystalline form of silicon dioxide (quartz) is hard and brittle.

### Question

26 Draw the Lewis structures for $SiO_4$ and $Si_2O$ fragments in quartz (figure 28). Estimate the value of the O–Si–O bond angle.

Above 1700°C, diamond turns into graphite, where the carbon atoms are arranged into flat layers with bond angles of 120°. Each carbon atom in graphite provides three electrons to form covalent bonds with its three immediate neighbours. The fourth electron is donated to a common electron cloud that surrounds all carbon atoms within the layer. These delocalized electrons can move freely, so graphite is a good conductor of electricity.

The separate layers of carbon atoms in graphite are held together in a stack by very weak London dispersion forces. As a result, these layers can slide past one another or be easily separated – for example, by pressing the tip of a graphite pencil to paper. Due to its softness, electrical conductivity and chemical inertness, graphite is commonly used as a lubricant and inert electrode material.

In addition to diamond and graphite, carbon forms several other *allotropes*, including graphene (a single layer of graphite), carbon nanotubes (graphene rolled into cylinders) and $C_{60}$ fullerene (figure 29).

**Key term**

Elementary substances with the same chemical composition but different structures are called **allotropes**.

**DP link**

The structures and properties of these allotropes will be discussed in detail in **4.3 Covalent structures** and **A Materials** of the IB Chemistry Diploma Programme.

## Question

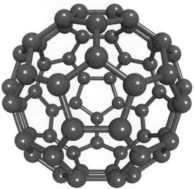

27 Phosphorus forms several allotropes. White phosphorus is composed of $P_4$ molecules, while red phosphorus contains long chains of covalently bonded atoms with no delocalized electrons. Compare and contrast the properties of these two allotropes in term of their melting and boiling points, volatility and electrical conductivity.

 **DP ready** | **Nature of science**

### Ozone

Molecular oxygen has two allotropes: dioxygen ($O_2$) and ozone ($O_3$). Under normal conditions, ozone is a gas with a characteristic pungent smell. People often describe this smell as "electric" because ozone is commonly produced from dioxygen by electric discharges:

$$3O_2 \rightarrow 2O_3$$

The molecule of ozone is V-shaped because of the presence of a lone electron pair on the central oxygen atom:

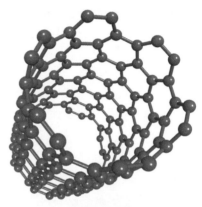

Figure 29. Molecular structures of a carbon nanotube (top) and $C_{60}$ fullerene (bottom)

Experimental data suggest that both O···O bonds in ozone have the same length and strength, so one of the shared electron pairs is delocalized between all three oxygen atoms. As a result, each O···O bond in ozone is intermediate between single and double.

Ozone is present in the Earth's stratosphere, where it absorbs a significant proportion of ultraviolet (UV) radiation from the Sun and thus protects all living organisms from its harmful effects. Elementary chlorine and its organic compounds destroy stratospheric ozone, so the release of such substances into the atmosphere must be minimized.

**Internal link**

For more about the mechanism by which elementary chlorine destroys ozone, see **5.2 Chemical kinetics**.

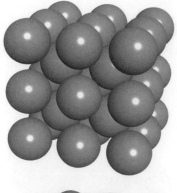

**Figure 30.** Metallic lattices of copper (top) and zinc (bottom)

## Metals and alloys

More than 85% of all elements in the periodic table are metals. Despite the differences in their properties, metals have many common characteristics, such as low ionization energy and low electronegativity. Because of that, atoms of metals readily lose their electrons and form positively charged ions (cations). In elementary substances, these cations adopt regular arrangements, known as *metallic lattices*, in which each ion has six to twelve immediate neighbours (figure 30). The electrostatic repulsion between cations is counteracted by delocalized electrons, which move freely through the whole sample of metal. As a result, metals are good conductors of electricity, both in the solid and liquid states.

Mechanical properties of metals can also be explained by the nature of metallic bonding. Delocalized electrons act not only as a "glue" for metal cations but also as a "grease", allowing the cations to slide past one another and reform the lattice after any mechanical deformation. Therefore, most metals are malleable (can be pressed or hammered into any shape without breaking or cracking) and ductile (can be drawn out into a thin wire). Metal alloys often have greater mechanical strength than individual metals, as the presence of different elements distorts the lattice and prevents the cations from sliding past one another.

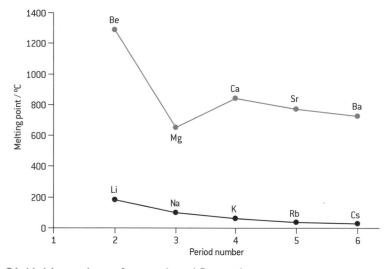

**Figure 31.** Melting points of group 1 and 2 metals

The strength of metallic bonding increases with ionic charge and decreases with ionic size. Smaller cations with higher charges experience greater electrostatic attraction to the cloud of delocalized electrons and thus produce stronger metallic lattices. Alkali metals have large atomic radii (figure 12) and form singly charged ions ($Li^+$ to $Cs^+$), so their melting points are among the lowest of all metals. As the atomic radii increase from lithium to caesium down the group, the melting points of these elements decrease (figure 31). Alkaline earth metals have slightly smaller atomic radii and form doubly charged ions ($Be^{2+}$ to $Ba^{2+}$), so they have much higher melting points than group 1 elements. With the exception of magnesium, melting points of alkaline earth metals also decrease down the group.

Transition metals tend to have very high melting points (typically between 1000 and 3000°C) and even higher boiling points (up to 5900°C for rhenium) because their atoms are capable of forming both metallic and covalent bonds with one another.

 **DP link**

The properties of transition metals will be discussed in **13 The periodic table—the transition metals** and **A Materials** in the IB Chemistry Diploma Programme.

## Chapter summary

In this chapter, you have learned about the electron structure of atoms and molecules, periodicity, and the nature of chemical bonding. Before moving on, please make sure that you have a working knowledge of the following concepts and definitions:

☐ Electrons in atoms can have only certain, quantized energies and occupy specific energy levels, also known as electron shells.

☐ Each energy level consists of one or more sublevels (s, p, d, f and so on), which in turn contain one or more atomic orbitals.

☐ Electrons in atomic orbitals form diffuse clouds of various shapes and sizes around the atomic nucleus.

☐ According to the Aufbau principle, electrons fill atomic orbitals in a certain order, starting from the lowest available energy levels and sublevels.

☐ Electron configurations of atoms can be represented by orbital diagrams (half-arrows in boxes) or shorthand notations, such as $1s^2 2s^2 2p^4$ or $[He]2s^2 2p^4$ for the oxygen atom.

☐ The properties of chemical elements depend on their electron configurations and are therefore periodic functions of the atomic number.

☐ In the periodic table, the elements are arranged in groups and periods according to their electron configurations.

☐ Certain groups of elements have common names, such as alkali metals (group 1), alkaline earth metals (group 2), chalcogens (group 16), halogens (group 17) and noble gases (group 18).

☐ Typical oxidation numbers for most elements can be deduced from their positions in the periodic table.

☐ Metals comprise over 85% of all elements and are located in the lower left part of the periodic table.

☐ Typical nonmetals (other than hydrogen) occupy the top right corner of the periodic table.

☐ Seven metalloids form a diagonal line from boron to astatine in the periodic table, separating metals from nonmetals.

☐ Metallic properties and atomic radii of the elements generally decrease across periods (left to right) and increase down groups.

☐ The type of chemical bonding depends on relative electronegativities of participating elements.

☐ Chemical bonds are classified as non-polar covalent, polar covalent, ionic and metallic.

☐ Intermolecular forces involve hydrogen bonds and van der Waals forces (dipole–dipole, dipole–induced dipole and London dispersion forces).

☐ Chemical and physical properties of substances depend on the nature and strength of chemical bonding.

☐ Ions in solid ionic compounds, metals and alloys form regular structures (lattices).

☐ Solid covalent compounds can have molecular or covalent network structures.

☐ Lewis structures can be used to predict the shapes of simple molecules using the VSEPR theory.

## Additional problems

1.  Read the article "Atomic orbital" on Wikipedia, paying particular attention to the energies, shapes and relative sizes of atomic orbitals with various $n$, $\ell$ and $m_\ell$ quantum numbers. Explore the orbital table by clicking on the images of specific orbitals. You may skip everything related to wave functions and relativistic effects.

2.  Using print and online resources, find at least one common technological application for each of the following elements: a) carbon; b) nitrogen; c) silicon; d) mercury; e) bromine; f) chromium; g) krypton.

3.  State the symbols and names for the following elements: a) the second d-element of the fourth period; b) the radioactive alkaline earth metal; c) the third member of the nitrogen group; d) the heaviest actinide; e) the noble gas of the sixth period.

4.  State electron configurations and draw orbital diagrams for the following species: a) sulfur atom; b) $S^{2-}$ anion; c) $Zn^{2+}$ cation.

5.  Determine the oxidation numbers of all elements in the following compounds: a) KI; b) $BaBr_2$; c) $NO_2$; d) $P_2O_5$; e) $CH_4$; f) $AlH_3$; g) $SO_3$.

6.  Outline the periodic trends in oxidation numbers and explain how these trends are related to the electron structure of the elements.

7.  Identify the bonding types in chemical compounds from question 5.

8.  Using partial charges, show the distribution of electron density in the molecules of ammonia and carbon tetrachloride.

9.  Identify the intermolecular forces in the following substances and arrange these substances in order of increasing melting points:
    a) $NH_3$, b) $N_2$, c) $H_2$, d) $H_2S$.

10. Draw the Lewis structures for the following species: a) $Cl_2$; b) $AlCl_3$; c) $NCl_3$; d) $OF_2$; e) $CF_4$; f) $NH_4^+$.

11. Deduce the shapes of molecules and ions from the previous problem and describe them in terms of electron domains and molecular geometry:

12. Boron trifluoride can form a coordinate bond with ammonia. Deduce the structure of the resulting molecule and the molecular geometries of boron and nitrogen centres.

# 3 Inorganic chemistry

> "Chemical nomenclature is ... a constant method of denomination, which helps the intelligence and relieves the memory."
>
> **Louis-Bernard Guyton de Morveau, *Observations sur la Physique, sur l'Histoire Naturelle et sur les Arts,* 1782, 19, 370–382.**

## Chapter context

In *1 Atomic theory and stoichiometry*, you learned that systematic names of binary compounds could be constructed from the names of **cations** and **anions**. You have also developed some basic skills in writing and balancing chemical equations involving elementary substances and **inorganic compounds**.

## Learning objectives

In this chapter you will learn about:
→ classification of **inorganic compounds**
→ their **nomenclature** and **chemical properties**
→ chemical reactions in terms of their **thermal effects, reversibility, spontaneity** and **number of participating species**
→ **balancing complex equations** with **changes in oxidation state**.

## 🔑 Key terms introduced

→ Oxides
→ Hydroxides
→ Acids and bases
→ Salts and acid salts
→ Exothermic, endothermic, spontaneous, non-spontaneous, reversible and irreversible reactions
→ Reduction, oxidation and redox reactions
→ Combination, decomposition, substitution and combustion reactions
→ Oxidizing and reducing agents
→ Activity series
→ Electrochemical processes
→ Voltaic cells
→ Anode and cathode
→ Cell diagrams
→ Electrolytic cells

## 3.1 Classification and nomenclature of inorganic compounds

You already know that *elementary substances* contain atoms of a single element while *chemical compounds* contain atoms of two or more elements bound together by chemical forces. Chemical compounds are traditionally divided into organic and inorganic. An *organic compound* must contain at least two elements, one of which is carbon. Any compound that does not contain carbon is considered to be *inorganic*. Simple derivatives of carbon, such as oxides (CO and $CO_2$), carbonic acid ($H_2CO_3$), carbonates and hydrogencarbonates (such as $Na_2CO_3$ and $NaHCO_3$), carbides (such as $Mg_2C$ and $Al_4C_3$) and cyanides (such as HCN and KCN) are also regarded as inorganic.

Inorganic compounds can be classified in several ways. For example, depending on the nature of their chemical bonding, inorganic compounds can be *ionic* (such as sodium chloride, NaCl) or *covalent* (such as carbon dioxide, $CO_2$).

Another type of classification is based on the number of elements in molecules or structural units of inorganic compounds. *Binary compounds*, such as NaCl and $CO_2$, are composed of two elements, while other compounds, such as $Na_2CO_3$ and $NaHCO_3$, consist of three or more elements.

Most inorganic compounds can be subdivided into four classes according to their chemical nature: oxides, hydroxides, acids and salts.

 **Internal link**

Elementary substances and chemical compounds are defined earlier in **1 Atomic theory and stoichiometry**.

The nomenclature and properties of organic compounds will be discussed in **7 Organic chemistry**.

 **Internal link**

The structures and properties of ionic compounds were discussed in **2.3 Chemical bonding**.

## Key term

**Oxides** are binary inorganic compounds containing oxygen.

## Oxides

*Oxides* are binary inorganic compounds containing oxygen. Since oxygen is very reactive, most oxides can be synthesized directly from elementary substances; for example:

$$2Ca(s) + O_2(g) \rightarrow 2CaO(s)$$
$$2H_2(g) + O_2(g) \rightarrow 2H_2O(l)$$
$$S(s) + O_2(g) \rightarrow SO_2(g)$$

Noble gases, halogens and some metals, such as silver, gold and platinum, do not react with molecular oxygen, but their oxides can be obtained indirectly. Compounds of oxygen with fluorine, $OF_2$ and $O_2F_2$, are not classified as oxides, as fluorine is more electronegative than oxygen.

Systematic names of oxides are formed by adding the word "oxide" to the element name. For example, CaO is calcium oxide and $H_2O$ is hydrogen oxide. If an element can form more than one oxide, the oxidation state of that element must be shown by Roman numerals in brackets after the element name; for example:

$SO_2$, sulfur(IV) oxide

$SO_3$, sulfur(VI) oxide

Alternatively, the number of each atom in the oxide can be stated explicitly, such as "sulfur dioxide" for $SO_2$ and "sulfur trioxide" for $SO_3$. Although both variants of nomenclature are accepted by IUPAC, the use of Roman numerals is preferred.

## Question

1 Complete the table below by deducing the formulae and/or names of oxides.

| Formula | Name | Alternative name |
|---------|------|------------------|
| FeO | | |
| | iron(III) oxide | |
| | | carbon monoxide |
| $CO_2$ | | |
| | nitrogen(I) oxide | |
| | | diphosphorus pentoxide |
| $Al_2O_3$ | | |

The oxidation state of oxygen in almost all oxides is −2. However, oxygen is also able to form *peroxides*, in which its oxidation state is −1:

$H_2O_2$, hydrogen peroxide

$BaO_2$, barium peroxide

Most peroxides are unstable and easily decompose to oxides by releasing excess oxygen:

$$2H_2O_2(l) \rightarrow 2H_2O(l) + O_2(g)$$
$$2BaO_2(s) \rightarrow 2BaO(s) + O_2(g)$$

Because of their unique chemical properties, peroxides are often considered as a separate class of inorganic compounds.

## Hydroxides

*Hydroxides*, or *metal hydroxides*, are inorganic compounds containing a metal atom and one or more hydroxyl groups, OH. Examples of metal hydroxides are:

NaOH, sodium hydroxide

$Fe(OH)_2$, iron(II) hydroxide

$Fe(OH)_3$, iron(III) hydroxide

Systematic names of hydroxides are constructed in the same way as the names of oxides, by combining the name of the metal with the word "hydroxide".

Depending on the electronegativity of the metal, the bonds between metal and OH groups can be covalent or ionic. Ionic hydroxides can be described as metal cations ($M^{n+}$) bonded to one or more hydroxide anions ($OH^-$).

When writing the chemical formulae of hydroxides, it is convenient to treat each OH group as single entity with an ionic charge of −1. In such a case, the number of OH groups in any metal hydroxide will be equal to the oxidation number of the metal atom. In the above examples, sodium (oxidation number +1) requires one OH group, iron(II) bonds to two OH groups, and so on. Similarly, if we know the formula of a metal hydroxide, we can always determine the oxidation state of the metal.

Hydroxides of alkali metals (group 1) and alkaline earth metals (group 2) are readily formed by the reactions of oxides with water, for example:

$$Na_2O(s) + H_2O(l) \rightarrow 2NaOH(aq)$$
$$CaO(s) + H_2O(l) \rightarrow Ca(OH)_2(aq)$$

 **Key term**

**Hydroxides** are inorganic compounds containing a metal atom and one or more hydroxyl groups, OH.

### Question

2   Write and balance the equations for the reactions of potassium oxide and barium oxide with water. State the systematic names of the reaction products.

Aqueous solutions of such hydroxides feel slippery to the touch, taste bitter and turn litmus (a natural pigment) blue. These characteristics are typical for *bases*. The chemical properties of metal hydroxides and other bases are opposite to those of *acids*, which are another common class of inorganic compounds.

**Internal link**

Bases are a diverse group of chemical compounds that are discussed in detail in **6 Acids and bases**.

In contrast to hydroxides of alkali and alkaline earth metals, most other hydroxides are insoluble in water and cannot be prepared directly from oxides. A table of the solubilities of metal hydroxides and other inorganic compounds is given in the appendix of this book.

At high temperatures, nearly all insoluble hydroxides decompose into oxides and water; for example:

$$Fe(OH)_2(s) \rightarrow FeO(s) + H_2O(g)$$
$$2Fe(OH)_3(s) \rightarrow Fe_2O_3(s) + 3H_2O(g)$$

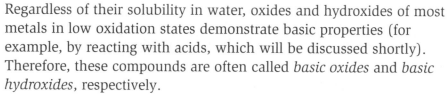

**Question**

3   Write and balance the equations for thermal decomposition of the following hydroxides: a) copper(I) hydroxide, b) copper(II) hydroxide, c) aluminium hydroxide.

Regardless of their solubility in water, oxides and hydroxides of most metals in low oxidation states demonstrate basic properties (for example, by reacting with acids, which will be discussed shortly). Therefore, these compounds are often called *basic oxides* and *basic hydroxides*, respectively.

Some nonmetallic ions can also form hydroxides. One such ion is ammonium, $NH_4^+$, which behaves as a single entity with a +1 ionic charge. Ammonium hydroxide, $NH_4OH$, is very unstable and exists only in aqueous solutions. Even at room temperature, ammonium hydroxide readily decomposes into ammonia and water:

$$NH_4OH(aq) \rightarrow NH_3(aq) + H_2O(l)$$

Ammonium is a very common ion in inorganic chemistry. To some extent, it resembles sodium, so if you know the formula of a sodium compound, you can always write the formula of its ammonium analogue by replacing Na with $NH_4$.

### Acids

*Acids* contain one or more hydrogen atoms that can be readily substituted with metals or other cations (such as ammonium). In *binary acids*, these hydrogen atoms are directly bonded to halogens or sulfur:

> HF, hydrogen fluoride
> HCl, hydrogen chloride
> HBr, hydrogen bromide
> HI, hydrogen iodide
> $H_2S$, hydrogen sulfide

Since hydrogen is the least electronegative nonmetal, its oxidation state in binary acids is always +1, while the other nonmetal has its lowest oxidation state (–1 for halogens or –2 for sulfur). Therefore, we can construct the formula of any binary acid in the same way as the formulae of other binary compounds.

**Figure 1.** Lemons taste sour due to the presence of organic acids

Another common acid, hydrogen cyanide (HCN), is a special case. Like the OH⁻ in hydroxides, the CN⁻ group in HCN behaves as a single entity with a formal charge of –1. Binary acids and hydrogen cyanide are collectively known as *hydroacids*.

*Oxoacids* usually consist of three elements, including hydrogen and oxygen (table 1).

**Table 1.** Common oxoacids and their anions

| Acid | | Anion | |
|---|---|---|---|
| $HNO_3$ | nitric | $NO_3^-$ | nitrate |
| $HNO_2$ | nitrous | $NO_2^-$ | nitrite |
| $H_2SO_4$ | sulfuric | $SO_4^{2-}$ | sulfate |
| $H_2SO_3$ | sulfurous | $SO_3^{2-}$ | sulfite |
| $H_3PO_4$ | phosphoric | $PO_4^{3-}$ | phosphate |
| $H_3PO_3$ | phosphorous | $PO_3^{3-}$ | phosphite |
| $HClO_4$ | perchloric | $ClO_4^-$ | perchlorate |
| $HClO_3$ | chloric | $ClO_3^-$ | chlorate |
| $HClO_2$ | chlorous | $ClO_2^-$ | chlorite |
| $HClO$ | hypochlorous | $ClO^-$ | hypochlorite |
| $H_2CO_3$ | carbonic | $CO_3^{2-}$ | carbonate |

 **Internal link**

Some more inorganic and organic acids are listed in table 1 of **6 Acids and bases**.

Systematic names of oxoacids consist of two words, the second of which is always "acid". The first word is derived from the name of the central element, such as "sulfur" in "sulfuric" acid ($H_2SO_4$) or "nitrogen" in "nitric acid" ($HNO_3$). The suffix "-ic" in these names indicates that the central element has its highest oxidation state (+6 for sulfur and +5 for nitrogen). Lower oxidation states of central elements in acids are usually designated by the suffix "-ous", such as in "sulfurous acid" ($H_2SO_3$) and "nitrous acid" ($HNO_2$), where sulfur and nitrogen have oxidation states of +4 and +3, respectively.

When an element forms oxoacids in three or four different oxidation states, the nomenclature becomes more complicated. The lowest oxidation state of the element is designated by the prefix "hypo-". The next two acids are named as usual, by adding suffixes "-ous" and "-ic" to the element name. Finally, the highest oxidation state of the central element is designated by the prefix "per-".

To determine the oxidation state of the central element, you need to remember that the oxidation states of hydrogen and oxygen in all oxoacids are +1 and –2, respectively. Therefore, you can find the oxidation state of the central element by adding together the oxidation states of all other atoms and reversing the sign.

**Worked example: Determining oxidation states in oxoacids**

1. Determine the oxidation state of nitrogen in nitric acid, $HNO_3$.

*Solution*

$HNO_3$ contains one H and three O atoms.

The sum of their oxidation states is $1 + 3 \times (-2) = -5$

So the remaining element (N) must have an oxidation state of $+5$ (as the sum of oxidation states of all elements in a neutral molecule must be equal to zero).

**Question**

4  Like chlorine, bromine forms four oxoacids of the general formula $HBrO_n$, where $n = 1$–$4$. Determine the oxidation states of bromine in these oxoacids and deduce their names using the rules described in the text.

To deduce the formula of an unknown oxoacid, we need to know the oxidation state of the third element (besides H and O) and the number of either H or O atoms in the acid molecule.

**Worked example: Deducing formulae using oxidation states**

2. Arsenic in its highest oxidation state forms an oxoacid with three hydrogen atoms. Deduce the formula of the acid.

*Solution*

Arsenic is a group 15 element, so its highest oxidation state is $+5$. If we write the formula of this acid as $H_3AsO_n$, the sum of oxidation states of H and As atoms will be $3 \times 1 + 5 = 8$.

To produce a neutral molecule, the sum of oxidation states of O atoms must be $-8$.

Therefore, $-2n = -8$ and so $n = 4$. The formula of the acid is $H_3AsO_4$.

**Question**

5  Deduce the formula of boric acid, which contains one boron and three oxygen atoms in its molecule.

Most oxoacids are readily formed from oxides of nonmetals and water:

$$SO_2(g) + H_2O(l) \rightarrow H_2SO_3(aq)$$
$$P_4O_{10}(s) + 6H_2O(l) \rightarrow 4H_3PO_4(aq)$$

Such oxides are known as *acidic oxides* (since they form acids) and also as *acid anhydrides* (from Greek *an-* "without" and *hudōr* "water"). In the above examples, sulfur(IV) oxide is the anhydride of sulfurous acid, and phosphorus(V) oxide is the anhydride of phosphoric acid.

**Question**

6  Write and balance the equations for the reactions of the following oxides with water:
a) carbon(IV) oxide; b) nitrogen(III) oxide; c) chlorine(VII) oxide.

Not all oxides of nonmetals are acidic oxides. For example, nitrogen(I) oxide ($N_2O$) and nitrogen(II) oxide (NO) do not react with water or form any acid indirectly. Such oxides are called *neutral* or *indifferent*.

As was mentioned earlier, chemical properties of acids are opposite to those of bases. Solutions of acids taste sour (like vinegar, which is an aqueous solution of ethanoic acid, $CH_3COOH$), turn litmus red and react with bases; for example:

$$HCl(aq) + NaOH(aq) \rightarrow NaCl(aq) + H_2O(l)$$
$$H_2SO_4(aq) + 2KOH(aq) \rightarrow K_2SO_4(aq) + 2H_2O(l)$$

The reactions between acids and bases are known as *neutralization* reactions. By neutralizing each other, the acid and the base lose their characteristic properties and produce two new compounds, one of which is water. The other compound, known as *salt*, consists of a metal cation and the acid anion (such as $Na^+$ and $Cl^-$ in $NaCl$). The nomenclature and properties of salts will be discussed later.

### Practical skills: Balancing acid–base equations

Stoichiometric coefficients in equations involving two chemical elements are easy to find by trial and error. Other equations, especially those involving four or more elements, can be a greater challenge. When balancing such equations, you can use the following strategy.

1. Balance all nonmetals except hydrogen and oxygen.

2. Balance all metals. If you need to change any stoichiometric coefficients deduced earlier, return to step 1.

3. Balance hydrogen. Again, if you need to change any coefficients, return to step 1.

4. At this point, the equation should be balanced already. To verify it, count oxygen atoms in the reactants and products. If their numbers do not match, return to step 1.

In most cases, this strategy produces a balanced equation with the fewest trials. Let's consider the following equation (the states are omitted for clarity):

$$H_3PO_4 + Ca(OH)_2 \rightarrow Ca_3(PO_4)_2 + H_2O \quad \text{(not balanced)}$$

Besides H and O, the equation involves only one nonmetal (P), which should be balanced first. There are two P atoms on the right but only one on the left, so we need to write a 2 in front of $H_3PO_4$:

$$2H_3PO_4 + Ca(OH)_2 \rightarrow Ca_3(PO_4)_2 + H_2O \quad \text{(not balanced)}$$

Step 1 is now complete, so we can look at metals. The only metal in the equation is Ca (one atom on the left, three atoms on the right), so we write 3 in front of $Ca(OH)_2$:

$$2H_3PO_4 + 3Ca(OH)_2 \rightarrow Ca_3(PO_4)_2 + H_2O \quad \text{(not balanced)}$$

Step 2 is complete, so the next element is hydrogen. On the left-hand side, there are $2 \times 3 = 6$ H atoms in $2H_3PO_4$ and $3 \times 2 = 6$ H atoms in $3Ca(OH)_2$, so the total number of H atoms in the reactants is $6 + 6 = 12$. On the right-hand side, there are only two H atoms in "$H_2O$", so the coefficient in front of $H_2O$ should be $12/2 = 6$:

$$2H_3PO_4 + 3Ca(OH)_2 \rightarrow Ca_3(PO_4)_2 + 6H_2O$$

Now we need to check the last remaining element, oxygen. There are $2 \times 4 = 8$ O atoms in $2H_3PO_4$ and $3 \times 2 = 6$ O atoms in $3Ca(OH)_2$, so we have $8 + 6 = 14$ O atoms on the left. On the right, there are $4 \times 2 = 8$ O atoms in $Ca_3(PO_4)_2$ and 6 O atoms in $6H_2O$, so the total number of O atoms on the right is also $8 + 6 = 14$. Therefore, oxygen is balanced, and so is the equation.

This strategy works well for acid–base and other equations where none of the elements change oxidation state. It can also be used for most reactions of acids with metals. The balancing of redox equations (those involving changes in the oxidation state of participating elements) will be discussed in *3.3 Redox processes*.

**Internal link**

The balancing of simple equations is described in **1.2 Chemical substances, formulae and equations**.

In neutralization reactions, neither acid anions ($Cl^-$ and $SO_4^{2-}$) nor metal cations ($Na^+$ and $K^+$) change their composition: the acid loses its $H^+$ cations, the base loses its $OH^-$ anions, and the remaining ions combine together into a salt. Therefore, the formulae of salts can be constructed exactly in the same manner as the formulae of binary compounds (in fact, NaCl and many other salts are indeed binary compounds).

For example, the salt formed in the reaction between $Fe(OH)_3$ and $HNO_3$ will contain the cation $Fe^{3+}$ and the anion $NO_3^-$. To balance the charges, we will need three $NO_3^-$ anions for each $Fe^{3+}$ cation, so the formula of the salt will be $Fe(NO_3)_3$.

## Question

**Q**

7 Write and balance the equations for the neutralization reactions between the following acids and bases: **a)** HCl(aq) and $Mg(OH)_2$(s); **b)** $H_2SO_4$(aq) and $Ca(OH)_2$(aq); **c)** $H_3PO_4$(aq) and KOH(aq); **d)** $H_2SO_4$(aq) and $Al(OH)_3$(s). Refer to the solubility table in the appendix to determine the states of salts formed in these reactions.

Most acids react with active metals, producing salts and molecular hydrogen:

$$2HCl(aq) + Fe(s) \rightarrow FeCl_2(aq) + H_2(g)$$
$$2H_3PO_4(aq) + 3Mg(s) \rightarrow Mg_3(PO_4)_2(s) + 3H_2(g)$$

These reactions can be used to discover the presence of acids in aqueous solutions: if the addition of magnesium or any other active metal produces bubbles of gas, the solution is likely to contain an acid. Please note that all alkali metals and most alkaline earth metals (calcium to radium) react violently with water, so they cannot be used for discovering acids in aqueous solutions.

Nitric acid and concentrated sulfuric acid also react with metals but do not produce hydrogen. These reactions will be discussed in *3.3 Redox processes*.

Acids can displace one another from their salts; for example:

$$Na_2CO_3(aq) + 2HCl(aq) \rightarrow 2NaCl(aq) + H_2CO_3(aq)$$

Carbonic acid is unstable and readily decomposes into carbon dioxide and water:

$$H_2CO_3(aq) \rightarrow CO_2(g) + H_2O(l)$$

Therefore, when an acid is added to a solution of sodium carbonate ($Na_2CO_3$) or sodium hydrogencarbonate ($NaHCO_3$), bubbles of carbon dioxide are released almost immediately:

$$Na_2CO_3(aq) + 2HCl(aq) \rightarrow 2NaCl(aq) + CO_2(g) + H_2O(l)$$
$$NaHCO_3(aq) + HCl(aq) \rightarrow NaCl(aq) + CO_2(g) + H_2O(l)$$

These reactions can also be used for discovering the presence of acids in solutions. In contrast to the reactions with metals, nitric and sulfuric acids react with carbonates and hydrogencarbonates in the same way as all other acids.

**DP ready** | **Theory of knowledge**

Chemical equations are often referred to as the language of chemistry. The use of a universal language helps communicate knowledge between scientists but at the same time limits our ability to describe natural phenomena that cannot be expressed in that language.

**Question**

8 Write and balance the equations for the following reactions: **a)** perchloric acid with magnesium; **b)** sulfuric acid with sodium carbonate; **c)** nitric acid with sodium hydrogencarbonate.

## Salts

Salts are ionic compounds composed of a metal (or ammonium) cation and an acid anion. Salts are commonly formed in the reactions between active metals or metal hydroxides with acids, as described before. Another method of producing salts is the reaction between metal oxides and acids, for example:

$$MgO(s) + 2HCl(aq) \rightarrow MgCl_2(aq) + H_2O(l)$$
$$Fe_2O_3(s) + 3H_2SO_4(aq) \rightarrow Fe_2(SO_4)_3(aq) + 3H_2O(l)$$

Ammonium salts can be formed either from unstable ammonium hydroxide ($NH_4OH$) or ammonia ($NH_3$):

$$NH_4OH(aq) + HCl(aq) \rightarrow NH_4Cl(aq) + H_2O(l)$$
$$NH_3(aq) + HCl(aq) \rightarrow NH_4Cl(aq)$$

Many acidic oxides, such as sulfur(IV) oxide, can also form salts by reacting with metal oxides or hydroxides:

$$2NaOH(aq) + SO_2(g) \rightarrow Na_2SO_3(aq) + H_2O(l)$$
$$CaO(s) + SO_2(g) \rightarrow CaSO_3(s)$$

Systematic names of inorganic salts are formed by combining the names of their cations and anions. For example, $CaSO_3$ contains the cation $Ca^{2+}$ ("calcium") and the anion $SO_3^{2-}$ ("sulfite", see table 1). Therefore, the name of this salt is "calcium sulfite".

> **Key term**
>
> **Salts** are ionic compounds composed of a metal (or ammonium) cation and an acid anion.

**Question**

9 Deduce the systematic names of the following salts: **a)** $Na_2SO_3$; **b)** $Al(NO_3)_3$; **c)** $NH_4Cl$; **d)** $Fe_2(SO_4)_3$; **e)** $Mg_3(PO_4)_2$. Refer to table 1 for the names of acid anions. Use Roman numerals to show the oxidation states of transition metals.

10 State the chemical formulae of the following salts: **a)** sodium nitrate; **b)** potassium phosphate; **c)** ammonium carbonate; **d)** iron(II) sulfate; e) chromium(III) nitrate.

*Acid salts* are products of incomplete neutralization of acids. One such salt, sodium hydrogencarbonate ($NaHCO_3$), is formed when one hydrogen atom in carbonic acid is substituted with a sodium ion:

$$H_2CO_3(aq) + NaOH(aq) \rightarrow NaHCO_3(aq) + H_2O(l)$$

Acids with three hydrogen atoms, such as phosphoric acid, can form two types of acid salt, with one or two hydrogen atoms in the acid anion:

$$H_3PO_4(aq) + NaOH(aq) \rightarrow NaH_2PO_4(aq) + H_2O(l)$$
$$H_3PO_4(aq) + 2NaOH(aq) \rightarrow Na_2HPO_4(aq) + 2H_2O(l)$$

> **Key term**
>
> **Acid salts** are salts that contain hydrogen and are able to undergo a further neutralization reaction.

Another common method of producing acid salts is the reaction of normal salts with acids:

$$2Na_3PO_4(aq) + H_3PO_4(aq) \rightarrow 3Na_2HPO_4(aq)$$
$$Na_3PO_4(aq) + 2H_3PO_4(aq) \rightarrow 3NaH_2PO_4(aq)$$

Systematic names of acids salts are constructed by adding the suffix "hydrogen-" to the name of the anion. For example, $Na_2HPO_4$ contains one H atom, so the name of this salt is "sodium hydrogenphosphate". Similarly, $NaH_2PO_4$ with two H atoms in the anion is called "sodium dihydrogenphosphate". Note that the prefix "hydrogen" (or "dihydrogen") and the anion name are written as a single word, without a space between them.

> **Question** Q
>
> 11  State the systematic names for the following acid salts: **a)** KHS; **b)** $NH_4HSO_3$; **c)** $Mg(HCO_3)_2$; **d)** $Ca(H_2PO_4)_2$.
>
> 12  Write and balance the equations for the formation of each acid salt from the previous problem.

**Internal link**

The possibility and extent of reactions involving acid salts depend on the relative strengths of the participating acids and bases, which are discussed in **4.3 Solutions of electrolytes** and **6.2 Classification and properties of acids and bases** in this book.

The chemical properties of acid salts are intermediate between those of normal salts and acids. For example, acid salts can react with bases to produce normal salts:

$$NaH_2PO_4(aq) + 2NaOH(aq) \rightarrow Na_3PO_4(aq) + 2H_2O(l)$$
$$Na_2HPO_4(aq) + NaOH(aq) \rightarrow Na_3PO_4(aq) + H_2O(l)$$

At the same time, acid salts can react with acids, producing new acids and salts:

$$NaH_2PO_4(aq) + HCl(aq) \rightarrow NaCl(aq) + H_3PO_4(aq)$$
$$Na_2HPO_4(aq) + H_3PO_4(aq) \rightarrow 2NaH_2PO_4(aq)$$

## 3.2 Classification of chemical reactions

### Classification methods

There are millions of known chemical substances that can participate in a vast number of chemical reactions. These reactions can be classified in many ways. One classification is based on the observable changes, such as heat released or consumed during a reaction. For example, coal (an impure form of carbon) readily burns in oxygen, producing carbon dioxide and a large amount of heat. Such reactions are called *exothermic*. The energy released in the form of heat is often shown in chemical equations by the symbol $Q$ with a positive sign:

$$C(s) + O_2(g) \rightarrow CO_2(g) + Q$$

In contrast, the reaction between nitrogen and oxygen is *endothermic*, as the formation of nitrogen(II) oxide consumes heat energy:

$$N_2(g) + O_2(g) \rightarrow 2NO(g) - Q$$

Many exothermic reactions are *spontaneous*, which means that they can proceed on their own and do not require any external source of energy. Once a piece of coal is set on fire, it will continue burning until all carbon is converted into carbon dioxide. Conversely, endothermic reactions are usually *nonspontaneous* and can proceed only when the energy is supplied from an external source. For example, nitrogen(II) oxide can be produced from molecular nitrogen and oxygen only at very high temperatures, when thermal energy is readily available.

However, some exothermic reactions are nonspontaneous, and certain endothermic reactions can proceed spontaneously (such as those used in instant cold packs).

Other observable chemical changes used to classify reactions include *precipitation* (formation of insoluble products) and *effervescence* (release of gases from liquids or solutions).

Chemical reactions can also be classified by their reversibility. *Irreversible* reactions proceed only in one direction, from reactants to products, and continue until at least one of the reactants is consumed completely. For example, if we mix together equal amounts of hydrogen and chlorine and initiate the reaction by heat or ultraviolet light, an explosion occurs, and both hydrogen and chlorine will be completely converted into hydrogen chloride:

$$H_2(g) + Cl_2(g) \rightarrow 2HCl(g)$$

In contrast, the reaction between gaseous hydrogen and iodine is *reversible* and proceeds in both directions, which is indicated by the equilibrium sign ($\rightleftharpoons$):

$$H_2(g) + I_2(g) \rightleftharpoons 2HI(g)$$

If we heat up a mixture of equal amounts of hydrogen and iodine, the reaction will never come to completion, and the mixture will always contain all three components ($H_2$, $I_2$ and HI). Similarly, if we heat up pure hydrogen iodide under the same conditions, it will partly decompose into hydrogen and iodine, and the final mixture will contain the same three components.

Chemical changes often affect the oxidation states of the elements involved. For example, in the reaction between hydrogen and chlorine, the more electronegative chlorine formally gains electrons and changes oxidation state from 0 in $Cl_2$ to –1 in HCl. In turn, hydrogen formally loses electrons and changes oxidation state from 0 in $H_2$ to +1 in HCl. A decrease in oxidation state (the gain of electrons) is known as *reduction* while an increase in oxidation state (the loss of electrons) is known as *oxidation*. Collectively, these processes are called *redox* reactions (from <u>red</u>uction and <u>ox</u>idation), and will be discussed later in this chapter. Non-redox processes include acid–base reactions and precipitation reactions.

 **Key term**

**Exothermic** reactions produce heat.

**Endothermic** reactions consume heat.

**Spontaneous** reactions are those that occur on their own, without an external source of energy.

 **Internal link**

Thermal effects and spontaneity of chemical reactions will be discussed in **5.1 Thermochemistry**.

Precipitation and effervescent reactions will be discussed in **4.3 Solutions of electrolytes**.

 **Internal link**

We will continue the discussion of reversible reactions in **5.3 Chemical equilibrium**.

 **Key term**

**Reduction** is a decrease in oxidation state, or the gain of electrons, and **oxidation** is an increase in oxidation state, or the loss of electrons. These two processes are collectively called **redox reactions**.

The most common method of classifying chemical reactions is based on the number of species involved in a chemical change. Using this approach, most inorganic reactions can be divided into four large groups: *combination, decomposition, substitution* and *exchange*.

## Combination reactions

*Combination reactions*, also known as *direct combination* or *synthesis reactions*, involve two reactants that form a single product. The general scheme of a *combination reaction* can be represented as follows:

$$A + B \rightarrow C$$

The reactants in combination reactions can be either elementary substances or chemical compounds, for example:

$$C(s) + O_2(g) \rightarrow CO_2(g)$$
$$2CO(g) + O_2(g) \rightarrow 2CO_2(g)$$
$$CaO(s) + CO_2(g) \rightarrow CaCO_3(s)$$

Notice that the numbers of reactants and products of a reaction refer to the types of participating species, not to their stoichiometric coefficients. For instance, the reaction between carbon monoxide and oxygen is classified as a combination reaction because it involves two types of reactant ($CO$ and $O_2$) and one type of product ($CO_2$), despite the fact that the balanced equation includes two molecules of $CO$ and two molecules of $CO_2$.

All combination reactions involving elementary substances are redox processes, as an elementary substance must change its oxidation state to form a compound. In contrast, nearly all combination reactions that involve only chemical compounds are non-redox processes, as illustrated by the last example.

## Decomposition reactions

*Decomposition reactions*, also known as *analysis reactions*, involve a single reactant that forms two or more products. The general scheme of a decomposition reaction is opposite to that of a combination reaction:

$$A \rightarrow B + C$$

Typical examples of decomposition reactions are:

$$2N_2O(g) \rightarrow 2N_2(g) + O_2(g)$$
$$H_2CO_3(aq) \rightarrow CO_2(g) + H_2O(l)$$
$$2KClO_3(s) \rightarrow 2KCl(s) + 3O_2(g)$$
$$NH_4HCO_3(s) \rightarrow NH_3(g) + CO_2(g) + H_2O(l)$$

Like combination reactions, any decomposition reaction that produces an elementary substance is a redox process while most reactions involving only chemical compounds are non-redox processes.

**Key term**

In **combination reactions**, two reactants combine to form a single product.

**Key term**

In **decomposition reactions**, a single reactant decomposes to form two or more products.

**Question**

13 Determine the oxidation states of all elements in the last four examples and verify that all chemical reactions involving elementary substances are redox processes.

## Substitution reactions

*Substitution reactions*, also known as *single displacement reactions*, involve two reactants, one of which is an elementary substance and another a chemical compound. The elementary substance displaces another elementary substance from the original compound and forms a new compound as follows:

$$A + BC \rightarrow AC + B$$

Reactions of active metals with acids are typical examples of substitution reactions:

$$Mg(s) + 2HCl(aq) \rightarrow MgCl_2(aq) + H_2(g)$$

Metals can also displace one another from their salts:

$$Fe(s) + CuSO_4(aq) \rightarrow FeSO_4(aq) + Cu(s)$$

Similarly, halogens and oxygen can substitute less electronegative nonmetals:

$$2KBr(aq) + Cl_2(g) \rightarrow 2KCl(aq) + Br_2(l)$$
$$2H_2S(g) + O_2(g) \rightarrow 2H_2O(l) + 2S(s)$$

All substitution reactions are redox processes, as they involve elementary substances.

## Exchange reactions

*Exchange reactions*, also known as *double displacement* or *metathesis reactions*, involve two chemical compounds as reactants and another two compounds as products:

$$AB + CD \rightarrow AD + CB$$

Many acid–base processes can be classified as exchange reactions, for example:

$$NaOH(aq) + HCl(aq) \rightarrow NaCl(aq) + H_2O(l)$$
$$NaHCO_3(aq) + NaOH(aq) \rightarrow Na_2CO_3(aq) + H_2O(l)$$
$$Na_2CO_3(aq) + H_2SO_4(aq) \rightarrow Na_2SO_4(aq) + H_2CO_3(aq)$$

Carbonic acid formed in the last equation is unstable and readily decomposes into carbon dioxide and water. Therefore, this equation can be written as follows:

$$Na_2CO_3(aq) + H_2SO_4(aq) \rightarrow Na_2SO_4(aq) + CO_2(g) + H_2O(l)$$

Such reactions are also classified as exchange processes, although they yield more than two reaction products.

The driving force of many exchange reactions is the formation of water or other stable covalent compounds. Other exchange reactions may involve precipitation; for example:

$$Na_2CO_3(aq) + CaCl_2(aq) \rightarrow CaCO_3(s) + 2NaCl(aq)$$
$$CuSO_4(aq) + 2NaOH(aq) \rightarrow Cu(OH)_2(s) + Na_2SO_4(aq)$$

 **Key term**

In **substitution reactions**, one elementary substance displaces another elementary substance from a compound to form a new compound.

 **Key term**

In **exchange reactions**, two chemical compounds react with each other to yield two different compounds as products.

 **Internal link**

Further discussion of exchange reactions requires the knowledge of the theory of electrolytic dissociation, which will be introduced in **4.3 Solutions of electrolytes**.

## Identifying the reaction type

The four most common types of inorganic reaction are summarized in table 2. Some redox processes, such as combustion, do not fit into any of these categories.

**Table 2.** Classification of inorganic reactions

| Reaction type | Number of reactants | Number of products | Reaction scheme |
|---|---|---|---|
| Combination | 2 or more | 1 | $A + B \rightarrow C$ |
| Decomposition | 1 | 2 or more | $A \rightarrow B + C$ |
| Substitution | 2 | 2 | $A + BC \rightarrow AC + B$ |
| Exchange | 2 | 2 or more | $AB + CD \rightarrow AD + CB$ |

### Question

14 Classify the following processes as combination, decomposition, substitution or exchange reactions:

a) $2NaI(aq) + Br_2(l) \rightarrow 2NaBr(aq) + I_2(s)$

b) $NaCl(aq) + AgNO_3(aq) \rightarrow AgCl(s) + NaNO_3(aq)$

c) $2NH_3(g) + H_2SO_4(aq) \rightarrow (NH_4)_2SO_4(aq)$

d) $2AgNO_3(aq) + H_2(g) \rightarrow 2Ag(s) + 2HNO_3(aq)$

e) $2KOH(aq) + H_3PO_4(aq) \rightarrow K_2HPO_4(aq) + 2H_2O(l)$

f) $4HNO_3(l) \rightarrow 2H_2O(l) + 4NO_2(g) + O_2(g)$

### Combustion

*Combustion reactions* involve two reactants, one of which is molecular oxygen. Since oxygen is abundant in nature and very reactive, combustion reactions are common, so it is convenient to consider them separately.

Combustion of all elementary substances and some binary compounds proceeds as a combination reaction, producing a single oxide; for example:

$$C(s) + O_2(g) \rightarrow CO_2(g)$$
$$2CO(g) + O_2(g) \rightarrow 2CO_2(g)$$

Combustion of other compounds proceeds as a substitution reaction, producing an elementary substance and an oxide:

$$4NH_3(g) + 3O_2(g) \rightarrow 2N_2(g) + 6H_2O(l)$$
$$4HCl(g) + O_2(g) \rightarrow 2Cl_2(g) + 2H_2O(l)$$

♾️ **Internal link**

Combustion of organic compounds will be discussed in **7.3 Properties of organic compounds**.

Some combustion reactions produce more than one oxide, so they do not resemble any of the above processes:

$$2H_2S(g) + 3O_2(g) \rightarrow 2SO_2(g) + 2H_2O(l)$$
$$4HCN(l) + 5O_2(g) \rightarrow 2N_2(g) + 4CO_2(g) + 2H_2O(l)$$

This type of combustion reaction is typical for organic compounds.

## 3.3 Redox processes

As was mentioned earlier, oxidation involves an increase in oxidation number (the loss of electrons) while reduction involves a decrease in oxidation number (the gain of electrons). Since the electrons are transferred from one atom or ion to another, every redox reaction must include at least two species, one of which undergoes oxidation and another undergoes reduction. For example, in the following reaction calcium is oxidized by losing its electrons to chlorine while chlorine is reduced by gaining electrons from calcium:

$$Ca(s) + Cl_2(g) \rightarrow CaCl_2(s)$$

The processes of oxidation and reduction can be represented by *electron half-equations* as follows, with the oxidation number as the superscript:

$$Ca^0 \rightarrow Ca^{+2} + 2e^-$$
$$Cl_2^0 + 2e^- \rightarrow 2Cl^{-1}$$

Electron half-equations provide a simple way of balancing redox reactions.

---

**Worked example: Balancing equations for redox reactions**

**3.** Deduce the balanced equation for the following process (the state symbols are omitted for clarity):

$$KClO_3 + H_2S \rightarrow KCl + S + H_2O \qquad \text{(not balanced)}$$

*Solution*

First of all, we need to determine which elements change oxidation states. Here, these elements are chlorine and sulfur. Chlorine changes its oxidation state from +5 in $KClO_3$ to −1 in $KCl$, so it undergoes reduction:

$$Cl^{+5} + 6e^- \rightarrow Cl^{-1}$$

Sulfur changes its oxidation state from −2 in $H_2S$ to 0 in S, so it undergoes oxidation:

$$S^{-2} \rightarrow S^0 + 2e^-$$

Since all reactants and products are electrically neutral, the total number of electrons lost by sulfur must be equal to the total number of electrons gained by chlorine. Therefore, for each $Cl^{+5}$ reduced to $Cl^{-1}$ there must be three $S^{-2}$ oxidized to $S^0$:

$$\begin{array}{l|l} Cl^{+5} + 6e^- \rightarrow Cl^{-1} & \times\ 1 \\ S^{-2} \rightarrow S^0 + 2e^- & \times\ 3 \end{array}$$

Thus to balance the electrons, we must place the coefficient 3 before $H_2S$ and S:

$$KClO_3 + 3H_2S \rightarrow KCl + 3S + H_2O \qquad \text{(not balanced)}$$

Note that the ratio between chlorine and sulfur is now fixed – we cannot change it without breaking the electron balance. If we ever need to change the coefficient before a substance involving Cl or S (for example, $KClO_3$), we will have to change all other coefficients (before $H_2S$, $KCl$ and S) proportionally, as otherwise the numbers of electrons gained and lost will not match.

To complete the balancing, we can follow the usual order of elements (first nonmetals except hydrogen and oxygen, then metals, then hydrogen and finally oxygen). The only metal is potassium, which is balanced already. To balance hydrogen, we have to place the coefficient 3 before the $H_2O$:

$$KClO_3 + 3H_2S \rightarrow KCl + 3S + 3H_2O$$

Finally, we need to check oxygen. There are three oxygen atoms on the left and three on the right, so the redox equation is now balanced.

**Internal link**

A similar method based on ionic half-equations is introduced in 4.3 Solutions of electrolytes.

The strategy in the worked example 3, known as the *oxidation number method*, works well for nearly all redox reactions.

**Question**

15 Balance the following equations using the oxidation number method:

a) $NH_3 + O_2 \rightarrow NO + H_2O$

b) $HI + H_2SO_4 \rightarrow H_2S + I_2 + H_2O$

c) $HBr + MnO_2 \rightarrow MnBr_2 + Br_2 + H_2O$

Many redox reactions involve elements in high oxidation states, such as $Cl^{+5}$ in $KClO_3$. These elements tend to gain electrons and undergo reduction. The compounds and ions containing such elements are often called *oxidizing agents*, as they cause other elements to lose electrons and undergo oxidation, while themselves being reduced. In addition to $KClO_3$, two common oxidizing agents are potassium permanganate ($KMnO_4$) and potassium dichromate ($K_2Cr_2O_7$), which contain transition metals in their highest oxidation states ($Mn^{+7}$ and $Cr^{+6}$, respectively). Other oxidizing agents that you might encounter are nitric acid ($HNO_3$), molecular halogens ($Cl_2$, $Br_2$ and $I_2$) and hydrogen peroxide ($H_2O_2$).

The elements in low oxidation states, such as $S^{-2}$ in $H_2S$, tend to lose electrons and undergo oxidation. Compounds and ions containing such elements are called *reducing agents*, as they are able to reduce other elements by giving them electrons. In addition to $H_2S$ and sulfides, common reducing agents are hydrogen halides ($HCl$, $HBr$ and $HI$) and their salts, sulfites (such as $Na_2SO_3$), nitrites ($NaNO_2$), compounds of iron(II) (such as $FeSO_4$) and many elementary substances, such as active metals.

**Key term**

**Oxidizing agents** oxidize other elements by causing them to lose electrons. **Reducing agents** reduce other elements by giving them electrons.

Deducing the products of redox reactions often requires the knowledge of inorganic chemistry beyond the level of the IB Chemistry Diploma Programme. Therefore, you will only be expected to construct simple redox equations based on the activity series, which is discussed later. However, you should be able to recognize common oxidizing and reducing agents, explain redox processes in terms of electron transfer, and balance redox equations with known reactants and products.

**Worked example: Balancing equations for redox reactions**

**4.** Consider the following unbalanced redox equation:

$K_2Cr_2O_7 + KI + H_2SO_4 \rightarrow K_2SO_4 + Cr_2(SO_4)_3 + I_2 + H_2O$

a) State the elements that undergo oxidation and reduction.

b) Identify the oxidizing and reducing agents.

c) Deduce the balanced equation for this redox process.

*Solution*

a) The oxidation state of chromium changes from $+6$ in $K_2Cr_2O_7$ to $+3$ in $Cr_2(SO_4)_3$, so chromium undergoes reduction (gains electrons).

   Iodine changes oxidation state from $-1$ in KI to 0 in $I_2$, so it undergoes oxidation (loses electrons).

**b)** Potassium dichromate ($K_2Cr_2O_7$) is the oxidizing agent and potassium iodide (KI) is the reducing agent.

**c)** First of all, we need to balance the electrons:

$$2Cr^{+6} + 6e^- \rightarrow 2Cr^{+3} \quad | \quad \times 1$$
$$2I^{-1} \rightarrow I_2^0 + 2e^- \quad | \quad \times 3$$

Now we can balance the oxidizing and reducing agents:

$$K_2Cr_2O_7 + 6KI + H_2SO_4 \rightarrow K_2SO_4 + Cr_2(SO_4)_3 + 3I_2 + H_2O$$

There are four sulfur atoms on the right, so let's write "4" before $H_2SO_4$ on the left:

$$K_2Cr_2O_7 + 6KI + 4H_2SO_4 \rightarrow K_2SO_4 + Cr_2(SO_4)_3 + 3I_2 + H_2O$$

There are eight potassium atoms on the left, so let's write "4" before $K_2SO_4$ on the right:

$$K_2Cr_2O_7 + 6KI + 4H_2SO_4 \rightarrow 4K_2SO_4 + Cr_2(SO_4)_3 + 3I_2 + H_2O$$

Unfortunately, this breaks the balance of sulfur (four atoms on the right, seven on the left), so we need to change the coefficient before $H_2SO_4$ from 4 to 7:

$$K_2Cr_2O_7 + 6KI + 7H_2SO_4 \rightarrow 4K_2SO_4 + Cr_2(SO_4)_3 + 3I_2 + H_2O$$

To balance hydrogen, we need to place "7" before water:

$$K_2Cr_2O_7 + 6KI + 7H_2SO_4 \rightarrow 4K_2SO_4 + Cr_2(SO_4)_3 + 3I_2 + 7H_2O$$

There are $7 + 7 \times 4 = 35$ oxygen atoms on the left and $4 \times 4 + 4 \times 3 + 7 = 35$ oxygen atoms on the right, so the equation is now balanced.

## Question

**16** Consider the following unbalanced redox equation:

$$KMnO_4 + KNO_2 + H_2SO_4 \rightarrow K_2SO_4 + MnSO_4 + KNO_3 + H_2O$$

**a)** State the elements that undergo oxidation and reduction.

**b)** Identify the oxidizing and reducing agents.

**c)** Deduce the balanced equation for this redox process.

## The activity series

Substitution reactions are a common type of redox process. As you learned in *3.2 Classification of chemical reactions*, halogens can displace each other from their salts; for example:

$$2KBr(aq) + Cl_2(g) \rightarrow 2KCl(aq) + Br_2(l)$$

In this reaction, molecular chlorine gains electrons and acts as an oxidizing agent while bromide ions lose electrons and act as a reducing agent:

$$Cl_2^0 + 2e^- \rightarrow 2Cl^{-1}$$
$$2Br^{-1} \rightarrow Br_2^0 + 2e^-$$

The oxidizing activity of halogens decreases down the group, along with their electronegativity. The electronegativity of chlorine ($\chi = 3.2$) is sufficient to displace both bromine ($\chi = 3.0$) and iodine ($\chi = 2.7$). Bromine can displace iodine but not chlorine, while iodine cannot displace any other halogens from their salts. Fluorine ($\chi = 4.0$) reacts with other halogens and water, so it is never used in substitution reactions.

## Question

**17** Complete and balance the following equations:

    **a)** $HBr(aq) + Cl_2(g) \rightarrow$ ...　　　　**b)** $NaI(aq) + Br_2(l) \rightarrow$ ...

## Internal link

Relative electronegativities of the elements are given in figure 15 in **2.2 The periodic law**.

## Key term

The **activity series** lists metals in order of their reactivity, with the strongest reducing agents being the most reactive metals.

The reducing activity of halides shows the opposite tendency to the oxidizing activity of halogens: it increases down the group. Hydrogen fluoride and its salts cannot be oxidized by chemical agents, as no element has a higher electronegativity than fluorine. Hydrogen chloride, hydrogen bromide and their salts react with strong oxidizing agents, such as potassium permanganate and potassium dichromate. Hydrogen iodide and its salts are oxidized very easily and can even react with molecular oxygen when exposed to air.

Like halogens, metals too can be arranged into the *activity series*, as shown in table 3. The first two rows of the table are occupied by alkali and alkaline earth metals, which are the most reactive elements and the strongest reducing agents. The least reactive metals and the weakest reducing agents, such as gold and platinum, are placed at the bottom of the table.

**Table 3.** The activity series of metals

| Metal | | | Ion | |
|---|---|---|---|---|
| Symbol | Reducing activity | Properties | Symbol | Oxidizing activity |
| Li, Na, K | | Displace hydrogen from water (Mg—slowly) | $Li^+$, $Na^+$, $K^+$ | |
| Ca, Sr, Ba | | | $Ca^{2+}$, $Sr^{2+}$, $Ba^{2+}$ | |
| Mg | | | $Mg^{2+}$ | |
| Al | | Protected by oxide | $Al^{3+}$ | |
| Zn | | Displace hydrogen from common acids | $Zn^{2+}$ | |
| Cr | | | $Cr^{3+}$ | |
| Fe | | | $Fe^{2+}$ | |
| Pb | | | $Pb^{2+}$ | |
| (H) | | Reference | $(H^+)$ | |
| Cu | | React with oxidizing agents | $Cu^{2+}$ | |
| Ag | | | $Ag^+$ | |
| Hg | | | $Hg^{2+}$ | |
| Au | | Very unreactive | $Au^{3+}$ | |
| Pt | | | $Pt^{4+}$ | |

Generally, a metal higher up in the activity series can displace any metal lower down from its salt or other compound. For example, a strip of zinc dipped into a solution of copper(II) sulfate quickly turns red because a layer of copper is deposited on its surface (figure 2):

$$Zn(s) + CuSO_4(aq) \rightarrow ZnSO_4(aq) + Cu(s)$$

In contrast, a piece of copper dipped into a solution of zinc(II) sulfate would remain unchanged, as copper is positioned below zinc in the activity series and thus cannot displace this more active metal from its salt:

$$Cu(s) + ZnSO_4(aq) \xrightarrow{\hspace{1cm}} \text{no reaction}$$

**Figure 2.** Reaction of zinc metal with a solution of copper(II) sulfate

## Question

18 Write and balance the equations for all possible substitution reactions between the following substances: Cr(s), Cu(s), Ag(s), $Cr(NO_3)_3$(aq), $Cu(NO_3)_2$(aq) and $AgNO_3$(aq).

The substitution of metals in aqueous solutions is possible only for the elements below aluminium, as more active metals will react with the water instead:

$$2Na(s) + 2H_2O(aq) \rightarrow 2NaOH(aq) + H_2(g)$$
$$Ca(s) + 2H_2O(aq) \rightarrow Ca(OH)_2(aq) + H_2(g)$$

Aluminium itself is very resistant to oxidation due to a thin protective film of oxide that quickly forms on its surface on contact with air. When this film is removed, aluminium tends to react with water rather than displace metals from their salts, so the use of aluminium as a reducing agent in aqueous solutions is impractical.

Metals from zinc to lead have moderate chemical activity. They do not react with water under normal conditions but displace hydrogen from nonoxidizing acids, such as hydrogen halides and dilute sulfuric acid:

$$Zn(s) + 2HCl(aq) \rightarrow ZnCl_2(aq) + H_2(g)$$
$$2Cr(s) + 3H_2SO_4(aq) \rightarrow Cr_2(SO_4)_3(aq) + 3H_2(g)$$

To emphasize this fact, hydrogen is also included in the activity series as a reference.

Metals from copper to mercury are very poor reducing agents and do not react with common acids. However, they can be oxidized by concentrated sulfuric and nitric acid as follows:

$$Cu(s) + 2H_2SO_4(conc.) \rightarrow CuSO_4(aq) + SO_2(g) + 2H_2O(l)$$
$$Ag(s) + 2HNO_3(conc.) \rightarrow AgNO_3(aq) + NO_2(g) + H_2O(l)$$

Dilute nitric acid also reacts with these metals, producing nitrogen(II) oxide:

$$3Cu(s) + 8HNO_3(aq) \rightarrow 3Cu(NO_3)_2(aq) + 2NO(g) + 4H_2O(l)$$
$$3Ag(s) + 4HNO_3(aq) \rightarrow 3AgNO_3(aq) + NO(g) + 2H_2O(l)$$

In contrast, dilute sulfuric acid has no effect on any metals that are less active than hydrogen.

---

### Question

19 Dilute nitric acid can be reduced by active metals to nitrogen oxides, molecular nitrogen or ammonium nitrate. Balance the following equations using the oxidation number method:

a) $Fe(s) + HNO_3(aq) \rightarrow Fe(NO_3)_3(aq) + NO(g) + H_2O(l)$

b) $Zn(s) + HNO_3(aq) \rightarrow Zn(NO_3)_2(aq) + N_2O(g) + H_2O(l)$

c) $Zn(s) + HNO_3(aq) \rightarrow Zn(NO_3)_2(aq) + N_2(g) + H_2O(l)$

d) $Mg(s) + HNO_3(aq) \rightarrow Mg(NO_3)_2(aq) + NH_4NO_3(aq) + H_2O(l)$

---

The last two metals in the activity series, gold and platinum, are extremely unreactive. They are resistant to almost all oxidizing agents, including concentrated sulfuric and nitric acids. Due to their inertness, these metals are commonly used as protective coatings for electrical connectors and laboratory equipment that must operate in aggressive chemical environment. In particular, platinum electrodes are used in many electrochemical experiments, which will be discussed in the final section of this chapter.

## Electrochemical processes: Voltaic cells

In a typical redox reaction, the electrons are transferred directly from one element to another. Figure 2 in the previous section shows an experiment in which zinc metal undergoes oxidation by passing its electrons to copper(II) sulfate, reducing $Cu^{2+}(aq)$ ions to copper metal. The sulfate ions remain unchanged, so they can be omitted from the equation:

$$Zn(s) + Cu^{2+}(aq) \rightarrow Zn^{2+}(aq) + Cu(s)$$

The oxidation of zinc metal and reduction of copper(II) ions can be represented by the following half-equations:

$$Zn(s) \rightarrow Zn^{2+}(aq) + 2e^- \text{ (oxidation)}$$
$$Cu^{2+}(aq) + 2e^- \rightarrow Cu(s) \text{ (reduction)}$$

In an alternative experimental setup, known as the *Daniell cell* (figure 3), the same half-reactions take place in different a, so the processes of oxidation and reduction are physically separated. The beaker on the left contains a zinc electrode submerged in a solution of zinc sulfate while the beaker on the right contains a copper electrode in a solution of copper(II) sulfate. Both electrodes are connected to a digital voltmeter that shows the difference in their electrical potentials. A strip of paper soaked in a solution of a salt, known as the *salt bridge*, completes the circuit by allowing ions to pass from one solution to another.

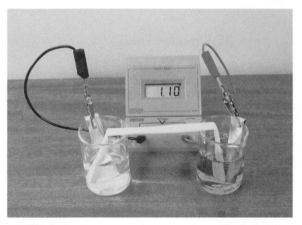

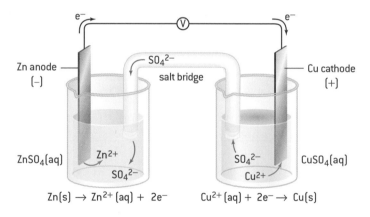

**Figure 3.** The Daniell cell

 **DP link**

You will learn more about the construction and function of electrochemical cells if you study **C.6 Electrochemistry, rechargeable batteries and fuel cells.**

When the cell is assembled, the material of the zinc electrode undergoes oxidation, producing $Zn^{2+}(aq)$ ions and electrons. These electrons flow through the external circuit to the copper electrode, where they reduce $Cu^{2+}(aq)$ ions to copper metal. The removal of $Cu^{2+}(aq)$ ions from the copper(II) sulfate solution leaves excess sulfate ions, which flow through the salt bridge to the left-hand side beaker and compensate the electric charge of newly formed $Zn^{2+}(aq)$ ions.

The Daniell cell demonstrates a spontaneous *electrochemical* process, in which the energy of chemical changes is converted into electrical energy. Electrochemical cells of this type are known as *voltaic cells*, as they produce electricity ("voltage") from spontaneous redox reactions. Voltaic cells are used in all electric batteries that supply energy to portable devices, electric vehicles and emergency power systems. Commercial voltaic cells use a variety of metals, such as lead (car batteries), nickel (NiCd and NiMH batteries) and lithium (Li-ion batteries).

Any electrochemical cell contains two electrodes, one of which is called the *anode* and another the *cathode*. The anode is the electrode at which oxidation takes place while the cathode is the electrode at which reduction takes place. In the Daniell cell, zinc acts as the anode and copper as the cathode.

In a voltaic cell, the anode has a negative charge, as the oxidation at this electrode proceeds spontaneously and produces excess electrons. Similarly, the reduction at the cathode consumes electrons, so the cathode in a voltaic cell is charged positively. Therefore, the electrons in the external circuit flow from the anode to the cathode.

The direction of ion flow through the salt bridge depends on the ionic charge. Anions, such as sulfate ions in the Daniell cell, travel from the cathode to the anode, as shown in figure 3. Cations flow in the opposite direction, from the anode to the cathode.

In chemical literature, voltaic cells are often represented by *cell diagrams*, or line notation. A typical diagram for the Daniell cell looks as follows:

$$Zn(s) \mid ZnSO_4(aq) \mid\mid CuSO_4(aq) \mid Cu(s)$$

In this diagram, all participating species are represented by their chemical symbols, the boundaries between the electrodes and solutions are shown as vertical lines, and the salt bridge as a double line. By convention, the anode (negative electrode) is shown on the left, and the cathode (positive electrode) is shown on the right. The ions that do not participate directly in the redox reactions (such as sulfate ions in the Daniell cell) can be omitted:

$$Zn(s) \mid Zn^{2+}(aq) \mid\mid Cu^{2+}(aq) \mid Cu(s)$$

The polarity of a simple voltaic cell can be determined using the activity series (table 3): the more active metal will undergo oxidation and thus act as the anode while the other metal will undergo reduction and act as the cathode.

 **Key term**

An **electrochemical** process involves interconversion of chemical and electrical energy.

**Voltaic cells** produce electricity from spontaneous redox reactions.

The **anode** is the electrode at which oxidation occurs, and the **cathode** is the electrode at which reduction occurs.

| DP ready | Approaches to learning |
|---|---|

**Cathode and anode**

To memorize the electrode names, you can use the acronym "**CROA**": **C**athode – **R**eduction; **A**node – **O**xidation.

Alternatively you may note that the words "cathode" and "reduction" both begin with a consonant while "anode" and "oxidation" begin with a vowel.

 **Key term**

A **cell diagram** represents a voltaic cell by chemical symbols and vertical lines. The anode is shown on the left, and the cathode on the right.

---

**Question**

20 Construct the cell diagram for a voltaic cell that involves the following redox reaction:

$$2AgNO_3(aq) + Cu(s) \rightarrow Cu(NO_3)_2(aq) + 2Ag(s)$$

State the names and signs of the electrodes. Deduce the direction of the electron flow in the external circuit and the ion flow in the salt bridge.

## Key term

**Electrolytic cells** use electrical energy to drive nonspontaneous reactions. Such processes are known as **electrolysis**.

## DP link

Electrode potentials are introduced in **19.1 Electrochemical cells** in the IB Chemistry Diploma Programme.

## Electrochemical processes: Electrolytic cells

Voltaic cells use the chemical energy of spontaneous redox reactions to generate electric current. The opposite process, known as *electrolysis*, takes place in *electrolytic cells*, where electricity is used as a source of energy for nonspontaneous reactions. For example, if we reverse the electric current in the Daniell cell (figure 3) by connecting its electrodes to an external power source, both the half-reactions will be reversed. The excess electrons at the zinc electrode will reduce $Zn^{2+}(aq)$ ions to zinc metal, while the material of the copper electrode will be oxidized to $Cu^{2+}(aq)$ ions:

$$Zn^{2+}(aq) + 2e^- \rightarrow Zn(s)$$
$$Cu(s) \rightarrow Cu^{2+}(aq) + 2e^-$$

However, the electrolysis of aqueous solutions is often complicated by the release of molecular hydrogen and oxygen from water:

$$2H_2O(l) \rightarrow 2H_2(g) + O_2(g)$$

As a result, both zinc and hydrogen will be formed at the cathode, while the reactions at the anode will produce copper(II) ions and oxygen. Further discussion of these processes requires knowledge of electrode potentials, so in this book, we will consider only the electrolysis of individual compounds, such as molten salts and oxides, where a single product is formed at each electrode.

The simplest electrolytic cell consists of a beaker filled with a molten salt, two electrodes and a power source (figure 4). The electrodes are usually made of platinum, graphite or other inert materials that can withstand high temperatures and aggressive chemical environment during the electrolysis. Since the electric current in an electrolytic cell is reversed with respect to the current in a voltaic cell, the electrodes change their signs: the anode becomes the positive electrode and the cathode becomes the negative electrode. However, oxidation still takes place at the anode and reduction at the cathode.

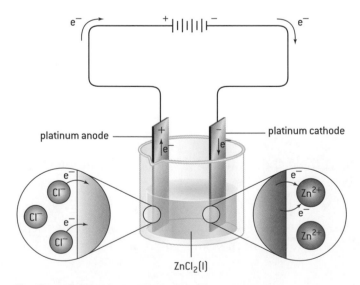

**Figure 4.** Electrolysis of molten zinc chloride

In the molten state, salts and other ionic compounds can conduct electricity owing to the presence of free-moving ions. The electrolytic cell in figure 4 uses molten zinc chloride, which is composed of

$Zn^{2+}$ cations and $Cl^-$ anions. During the electrolysis, zinc cations move to the negative electrode (cathode) and gain electrons, producing zinc metal:

$$Zn^{2+} + 2e^- \rightarrow Zn(l)$$

At the same time, chloride anions move to the positive electrode (anode) and lose electrons, producing molecular chlorine:

$$2Cl^- \rightarrow Cl_2(g) + 2e^-$$

Liquid zinc is collected at the bottom of the beaker while gaseous chlorine escapes from the reaction mixture. The overall redox process is represented by the following equation:

$$ZnCl_2(l) \rightarrow Zn(l) + Cl_2(g)$$

**Question**

21 Write and balance the equation for the electrolysis of molten sodium chloride. Explain the processes that take place at the electrodes in terms of redox half-equations and the flow of electrons and ions in the electrolytic cell.

Other electrolytic cells may have a different construction but operate on the same basic principle. For example, aluminium metal is produced by the electrolysis of aluminium oxide, $Al_2O_3$, as shown in figure 5. Since the melting point of pure $Al_2O_3$ is very high (over 2000°C), it is mixed with molten cryolite ($Na_3AlF_6$). This mixture remains liquid below 1000°C, which makes the process less energy-hungry and less expensive in terms of equipment.

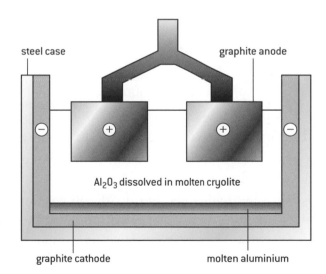

steel case    graphite anode

$Al_2O_3$ dissolved in molten cryolite

graphite cathode    molten aluminium

**Figure 5.** Electrolysis of aluminium oxide

The electrolysis takes place in a large steel container lined with graphite that acts as the cathode. The anode, also made of graphite, is lowered into the reaction mixture from above. During the electrolysis, aluminium cations are reduced at the cathode and form a layer of liquid aluminium at the bottom of the container:

$$Al^{3+} + 3e^- \rightarrow Al(l)$$

The oxygen anions are oxidized at the anode, producing molecular oxygen:

$$2O^{2-} \rightarrow O_2(g) + 4e^-$$

**DP link**

The production of aluminium and other metals is discussed in greater detail in **A.2 Metals and inductively coupled plasma (ICP) spectroscopy** in the IB Chemistry Diploma Programme.

Therefore, the balanced equation for the electrolysis of aluminium oxide looks as follows:

$$2Al_2O_3(l) \rightarrow 4Al(l) + 3O_2(g)$$

However, nearly all oxygen formed at the anode immediately reacts with the graphite:

$$O_2(g) + C(s) \rightarrow CO_2(g)$$

Because of that, the graphite anode is gradually consumed, so it has to be replaced from time to time.

The overall process in the electrolytic cell is a combination of the last two equations:

$$2Al_2O_3(l) + 3C(s) \rightarrow 4Al(l) + 3CO_2(g)$$

## Chapter summary

In this chapter, you have learned about different classes of inorganic compounds and their chemical changes, including redox reactions and electrochemistry. Before moving further, please make sure that you have a working knowledge of the following concepts and definitions:

- ☐ Inorganic compounds can be classified by the type of bonding, elemental composition and chemical nature.
- ☐ Systematic names of all compounds except oxoacids are constructed in the same way as the names of binary ionic compounds.
- ☐ Names of oxoacids and their anions are derived from element names by adding suffixes and prefixes.
- ☐ Variable oxidation states are shown by Roman numerals in parentheses after the element names.
- ☐ The four common classes of inorganic compounds are oxides, hydroxides, acids and salts.
- ☐ Acids and bases have opposite chemical properties and can neutralize one another, producing salts and water.
- ☐ Acid salts are products of incomplete neutralization of acids.
- ☐ Most acids react with active metals, producing salts and molecular hydrogen.
- ☐ Nitric and concentrated sulfuric acid do not produce hydrogen in reactions with metals.
- ☐ Chemical reactions can be classified by their spontaneity, thermal effect, reversibility, changes in the oxidation state, and the number of participating species.
- ☐ Common reaction types are combination (synthesis), decomposition (analysis), substitution (single displacement), exchange (double displacement) and combustion.
- ☐ Redox processes involve reduction (gain of electrons) and oxidation (loss of electrons).
- ☐ Any chemical reaction involving an elementary substance is a redox process.
- ☐ Redox equations can be balanced using the oxidation number method.
- ☐ Many oxidizing agents contain elements in high oxidation states, such as $Mn^{+7}$ in $KMnO_4$, $Cr^{+6}$ in $K_2Cr_2O_7$ and $N^{+5}$ in $HNO_3$.
- ☐ Other oxidizing agents are halogens, molecular oxygen and hydrogen peroxide.
- ☐ Reducing agents often contain elements in low oxidation states, such as $S^{-2}$ in $H_2S$, $I^{-1}$ in $HI$ and $Fe^{+2}$ in $FeSO_4$.
- ☐ Active metals and some other elementary substances can also act as reducing agents.
- ☐ Metals and halogens can be arranged into activity series.
- ☐ A metal higher up in the activity series can displace any metal lower down from its salt or other compound.

☐ Metals above zinc in the activity series react with water, so they cannot displace other metals in aqueous solutions.

☐ Metals below hydrogen in the activity series react with nitric acid and concentrated sulfuric acid, producing oxides of nitrogen and sulfur dioxide, respectively.

☐ Electrochemical processes involve chemical changes that release or consume energy in the form of electricity.

☐ Voltaic cells use spontaneous redox reactions to produce electric energy while electrolytic cells use electricity as a source of energy for nonspontaneous redox reactions.

☐ In all electrochemical cells, reduction occurs at the cathode and oxidation at the anode.

☐ In a voltaic cell, the cathode is the positive electrode and the anode is the negative electrode.

☐ In an electrolytic cell, the cathode is the negative electrode and the anode is the positive electrode.

☐ The salt bridge allows the movement of ions between half-cells and completes the electric circuit.

☐ Voltaic cells are often represented by cell diagrams, in which the anode is shown on the left.

☐ Electrolysis of a molten salt or oxide produces a metal at the cathode and a nonmetal at the anode.

## Additional problems

1. Describe the following chemical substances in terms of chemical bonding, elemental composition and chemical nature: a) $KBr$; b) $Br_2$; c) $K_2O$; d) $KHSO_3$; e) $H_2SO_4$.

2. State the systematic names for all compounds from the previous question.

3. Write the chemical formulae of the following compounds: a) nitric acid; b) hydrogen sulfide; c) ammonium hydrogensulfide; d) calcium hydrogencarbonate; e) manganese(IV) oxide; f) magnesium peroxide; g) copper(II) hydroxide; h) iron(III) sulfate.

4. Copper(II) sulfate, $CuSO_4$, was known for centuries as "blue vitriol" (from Latin *vitrum* "glass") due to its ability to form large, clear, deep-blue crystals. The development of systematic nomenclature made this and many other trivial names obsolete. Discuss what has been lost and gained in this process.

5. Write and balance as many chemical equations as you can using the following reactants: sodium oxide, phosphoric acid, sulfur(VI) oxide and water.

6. Describe, in your own words, how the presence of acids and bases in aqueous solutions can be discovered.

7. Without looking at the text, give one example of each common reaction type (addition, decomposition, substitution, exchange and combustion).

8. Classify the following reactions in terms of their thermal effect, reversibility, number of participating species and change in oxidation state of the elements.

   a) $2H_2O_2(aq) \rightarrow 2H_2O(l) + O_2(g) + Q$

   b) $Zn(s) + H_2SO_4(aq) \rightarrow ZnSO_4(aq) + H_2(g) + Q$

   c) $2AgNO_3(aq) + CuCl_2(aq) \rightarrow Cu(NO_3)_2(aq) + 2AgCl(s) + Q$

   d) $2NH_3(g) \rightleftharpoons N_2(g) + 3H_2(g) - Q$

   e) $2NO(g) + O_2(g) \rightleftharpoons 2NO_2(g) + Q$

   f) $Ca(OH)_2(s) + 2HNO_3(aq) \rightarrow Ca(NO_3)_2(aq) + 2H_2O(l) + Q$

9. Molecular chlorine reacts with a hot solution of potassium hydroxide to produce potassium chloride, potassium chlorate and water. Write and balance the equation for this reaction using the oxidation number method. Refer to table 1 for the formulae of acid anions.

10. Balance the following redox equations using the oxidation number method:

a) $S(s) + HNO_3(conc.) \rightarrow H_2SO_4(aq) + NO_2(g) + H_2O(l)$

b) $Zn(s) + H_2SO_4(conc.) \rightarrow ZnSO_4(aq) + S(s) + H_2O(l)$

c) $Mg(s) + H_2SO_4(conc.) \rightarrow MgSO_4(aq) + H_2S(g) + H_2O(l)$

d) $KMnO_4(aq) + HCl(aq) \rightarrow KCl(aq) + MnCl_2(aq) + Cl_2(g) + H_2O(l)$

e) $FeCl_2(aq) + H_2O_2(aq) + HCl(aq) \rightarrow FeCl_3(aq) + H_2O(l)$

11. Determine the volume of nitrogen(II) oxide produced at STP when 10.0 g of iron metal reacted with excess dilute nitric acid.

The unbalanced equation of this reaction is given in question 19 above (*3.3 Redox processes*), and the definition of STP is given in *1.3 Stoichiometric relationships*.

12. Using table 3, state two metals that can displace iron from an aqueous solution of iron(II) sulfate. Write and balance the equations for these substitution reactions.

13. Draw the cell diagram for a voltaic cell that contains the following substances: Cu(s), Cr(s), $CuCl_2(aq)$ and $CrCl_3(aq)$. State the names and signs of the electrodes, explain the function of the salt bridge and deduce the direction of the electron and ion flow.

14. Write and balance the equations for the electrolysis of the following molten salts: a) sodium chloride; b) lead(II) bromide; c) lithium sulfide.

15. An industrial electrolytic cell produces 15 kg of aluminium metal per hour and uses a block of graphite as the anode.

a) Assuming that all oxygen formed in the cell reacts with the anode material, calculate the mass of graphite consumed every hour.

b) Determine the time until the anode has to be replaced if its initial mass is 75 kg.

c) A small percentage of carbon dioxide formed in the cell reacts with molten aluminium to produce aluminium oxide and carbon monoxide. Deduce the balanced equation for this reaction and suggest how this process affects the efficiency of the electrolysis.

d) Suggest why the anode is made of graphite rather than platinum or other noble metal.

# Solutions and concentration

*In a great number of the cosmogonic myths the world is said to have developed from a great water, which was the prime matter. In many cases, ... this prime matter is indicated as a solution, out of which the solid earth crystallized out.*

**Svante Arrhenius, *Theories of Solutions* (1912)**

## Chapter context

Many chemical reactions are carried out in **solutions**. Sometimes it is only a matter of convenience, as solutions are easier to handle and mix than solids or gases. At the same time, the **solvent** can affect the reaction's direction, rate and extent.

## Learning objectives

In this chapter you will learn about:

→ different types of **solution**

→ the **quantitative composition** of solutions

→ the theory of **electrolytic dissociation**

→ the reactions of **electrolytes** in aqueous **solutions**

→ ionic equations.

## Key terms introduced

→ Homogeneous and heterogeneous mixtures

→ Solvents, solutes and solutions

→ Solubility; soluble, sparingly soluble and insoluble substances

→ Saturated and unsaturated solutions

→ Concentrated and dilute solutions

→ Standard ambient temperature and pressure (SATP)

→ Mass percentage ($\omega\%$) and mass fraction

→ Molar concentration ($c$)

→ Titrations and standard solutions

→ Strong and weak electrolytes

→ Degree of ionization ($\alpha$)

→ Ionic equations; total and net ionic equations

→ Spectator ions

→ Precipitation reactions and effervescent reactions

## 4.1 Composition and classification of solutions

### Types of solution

*Solutions* are homogeneous mixtures of two or more components. The term *homogeneous* means that the components of the mixture are uniformly distributed throughout the whole volume of the solution. For example, if we dissolve a tablespoon of sugar in a glass of water, each crystal of the sugar will break down into individual molecules, and these molecules will spread between the molecules of water in such a way that the resulting solution (syrup) will have exactly the same composition everywhere.

In contrast, if we stir a tablespoon of oil into a glass of water, the resulting mixture will not be completely uniform but will consist of small droplets of oil suspended in water. This *heterogeneous* mixture has a variable composition: oil inside each droplet and water outside. Heterogeneous mixtures are not classified as solutions, so we will not discuss them in this chapter.

Each solution consists of a *solvent* and one or more *solutes*. The solvent is usually the major component of the solution, so the properties of the whole solution are more or less similar to the properties of the solvent. Other components of the solution are called solutes. For example, a solution of sugar in water is more similar to water (clear colourless liquid) than sugar (white crystalline powder), so water is the solvent while sugar is the solute.

### Key term

A **homogeneous** mixture comprises uniformly distributed components, whereas in a **heterogeneous** mixture the components are distributed unevenly.

### Internal link

More examples of homogeneous and heterogeneous mixtures are given in **1.2 Chemical substances, formulae and equations**.

**Key term**

A **solvent** is the liquid that dissolves the solute.

A **solute** is the substance that dissolves in the solvent to make a solution.

A **solution** is a homogeneous, usually liquid, mixture containing two or more substances.

sugar
*solute*

+

water
*solvent*

=

syrup
*solution*

**Key term**

An **aqueous** solution is a solution in water.

**Key term**

**Miscible** liquids form a solution when added together in any proportion.

A **saturated solution** cannot dissolve any more solute.

An **unsaturated solution** can dissolve more solute.

**Key term**

**Solubility** is the maximum quantity of the solute that can be dissolved in a given quantity of the solvent.

In some cases, the identity of the solvent is unclear: for example, if we mix ethanol and water, each of these liquids can be called a solvent. However, if water is present in the mixture, it is traditionally regarded as the solvent, even if it is not the major component. In this example, we say "96% solution of ethanol in water" rather than "4% solution of water in ethanol".

Solutions in water are called *aqueous solutions* (from Latin *aqua*, water).

Organic liquids, such as ethanol, can also be used as solvents. For example, sugar can be dissolved in ethanol instead of water. Solutions that do not contain water are called *nonaqueous solutions*.

> **Question**
>
> 1 Classify the following mixtures by their composition and state:
> a) sea water; b) soda water; c) air; d) petrol (gasoline);
> e) fine sand shaken with water.
> Try to use as many characteristics as possible. For example, hydrochloric acid is an aqueous solution of a gaseous solute (HCl) in a liquid solvent ($H_2O$).

### Solubility

Some liquids, like water and ethanol, are *miscible*: they form homogeneous mixtures in any proportion. Other substances can produce solutions only when they are mixed in certain ratios. For example, if we stir 100 g of water with 100 g of sodium chloride at room temperature, only 36 g of the salt will dissolve in the water. The remaining 64 g will stay as a solid at the bottom of the vessel. The resulting solution is *saturated*, as no more salt can be dissolved in it. In contrast, an *unsaturated* solution of sodium chloride – one containing less than 36 g of sodium chloride per 100 g of water – can dissolve some more salt, until it becomes saturated.

This mass ratio (36 g:100 g) is known as the *solubility* of sodium chloride in water. Generally, the solubility of a solute is the ratio between the mass or volume of this substance and the mass or volume of the solvent in a saturated solution. The solubilities of solids are usually expressed as mass ratios (table 1). For liquids and gases, volume ratios can also be used.

**Table 1.** Solubility of salts, acids and bases in water at 25°C

| Compound | Solubility / g per 100 g | Compound | Solubility / g per 100 g |
|----------|--------------------------|----------|--------------------------|
| AgCl | 0.000191 | HF | miscible |
| $CaSO_4$ | 0.205 | HCl | 69.7 |
| NaCl | 36.0 | $H_2SO_4$ | miscible |
| KBr | 68.0 | $NH_3$ | 31.6 |
| $NaNO_3$ | 91.6 | NaOH | 113 |
| $NH_4NO_3$ | 217 | $Ca(OH)_2$ | 0.167 |
| $ZnBr_2$ | 482 | $Ba(OH)_2$ | 4.68 |

Informally, a substance is called:

- *soluble* if its aqueous solubility is over 1 g per 100 g of water
- *slightly soluble* or *sparingly soluble* if its solubility is between 0.01 and 1 g per 100 g of water
- *insoluble* if its solubility is below 0.01 g per 100 g of water.

The term "insoluble" does not suggest that a substance is absolutely insoluble in water; it only means that the solubility of that compound is so low that it can be ignored for most practical purposes.

### Question

2   Classify the compounds listed in table 1 as "soluble", "sparingly soluble" or "insoluble".

It is important to distinguish between the concepts of saturated/unsaturated and concentrated/dilute solutions.

- When we say that a solution is *concentrated*, we mean that the dissolved substance constitutes a significant proportion of the solution (i.e., the solute to solvent ratio is high).
- In contrast, a *dilute* solution contains very little solute with respect to the solvent.

There is no direct relationship between saturation and concentration: a saturated solution can be very dilute while an unsaturated solution can be highly concentrated (table 2).

**Table 2.** Saturation and concentration

| Solution | Unsaturated | Saturated |
|---|---|---|
| **dilute** | 1 g of NaCl in 100 g of $H_2O$ | 0.000191 g of AgCl in 100 g of $H_2O$ |
| **concentrated** | 200 g of $ZnBr_2$ in 100 g of $H_2O$ | 36 g of NaCl in 100 g of $H_2O$ |

The terms "concentrated" and "dilute" are not precisely defined and should be used with care. For example, most chemists would call a solution of 5 g of sulfuric acid in 100 g of water "dilute", as much higher proportions of sulfuric acid in water are often used in laboratories. At the same time, a solution of 5 g of potassium permanganate in 100 g of water would be considered very concentrated by any medical worker, as typical concentrations of potassium permanganate in antiseptic solutions are less than 0.1 g per 100 g of water.

Generally, the term:

- "concentrated" refers to solutions with much more than 10 g of the solute per 100 g of the solvent;
- "dilute" refers to solutions with much less than 10 g of the solute per 100 g of the solvent.

### Question

3   Using the data from table 1, classify the following solutions as saturated/unsaturated and dilute/concentrated:
  a) 56.5 g of NaOH in 50 g of $H_2O$;
  b) 2.05 g of $CaSO_4$ in 10 kg of $H_2O$;
  c) 11.7 g of $Ba(OH)_2$ in 250 g of $H_2O$. All data are given for 25°C.

### Internal link

A table of the solubility of inorganic compounds is given in the appendix of this book. The table is particularly useful for deducing ionic equations, which will be discussed at the end of this chapter.

### Key term

A **concentrated** solution has a high ratio of solute to solvent.

A **dilute** solution has a low ratio of solute to solvent.

*Temperature* and *pressure* affect the solubilities of solids, liquids and gases in different ways.

- For most solids, the solubility in water increases with temperature (figure 1).
- In contrast, most gases become less soluble at higher temperatures.
- The solubility of liquids depends on their nature; it can increase or decrease with temperature.

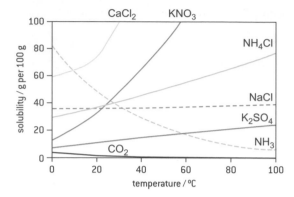

**Figure 1.** The effect of temperature on solubility

- The solubility of a gas (expressed in grams per 100 g of water) is roughly proportional to its pressure.
- Solids and liquids are incompressible, so pressure has almost no effect on their solubilities.

---

**DP ready** **Nature of science**

**Decompression sickness**

When you open a bottle of soda water, the pressure of carbon dioxide in the bottle decreases and some dissolved gas fizzes out. A similar process takes place in the body of a scuba diver who returns to the surface too rapidly: the nitrogen dissolved in the blood and other tissues produces bubbles, causing severe pain. This condition, decompression sickness, is also known as the bends or caisson disease, as it was first noted in workers constructing tunnels in caissons (temporary underwater structures).

---

Common units of solubility (g per 100 g) are useful when we need to prepare a saturated solution of a certain substance. However, when the solution is already prepared and we need to work out how much to use, other ways of expressing the quantitative composition of solutions are more useful.

## 4.2 Concentration expressions and stoichiometry
### Mass percentage

The most common way of expressing the concentration of a solution is its percent composition. The *mass percentage* ($\omega$%) of the solute is the ratio of its mass to the mass of the whole solution, expressed as a fraction of 100 (per cent):

$$\omega\% = \frac{m_{solute}}{m_{solution}} \times 100\%$$

A similar quantity, *mass fraction* ($\omega$), is the simple ratio of the same masses:

$$\omega = \frac{m_{solute}}{m_{solution}}$$

From the above expressions, it is obvious that $\omega\% = \omega \times 100\%$.

---

**Worked example: Calculating mass percentages**

**1.** Calculate the mass percentages of sodium chloride and water in a saturated solution that contains 36 g of sodium chloride in 100 g of water.

*Solution*

$m(\text{solution}) = m(\text{NaCl}) + m(\text{H}_2\text{O}) = 36 + 100 = 136\text{ g}$

$\quad\omega(\text{NaCl}) = m(\text{NaCl})/m(\text{solution}) = 36/136 \approx 0.265$

$\quad\omega\%(\text{NaCl}) = \omega(\text{NaCl}) \times 100\% = 0.265 \times 100\% = 26.5\%.$

To find the mass percentage of water, we can simply subtract the mass percentage of sodium chloride from 100%:

$\omega\%(\text{H}_2\text{O}) = 100\% - 26.5\% = 73.5\%$

Alternatively, we can get the same answer by dividing the mass of water by the mass of the solution:

$$\omega\%(\text{H}_2\text{O}) = \frac{m(\text{H}_2\text{O})}{m(\text{solution})} \times 100\% = \frac{100\text{ g}}{136\text{ g}} \times 100\% \approx 73.5\%$$

---

Note that the terms "mass percentage" and "mass fraction" must refer to specific compounds, not the whole solution. For example, it is incorrect to say that "the mass percentage of a saturated sodium chloride solution is 26.5%", as it is not clear whether we are talking about the percentage of sodium chloride or water. The correct statement would be "the mass percentage of sodium chloride in its saturated solution is 26.5%".

**Question**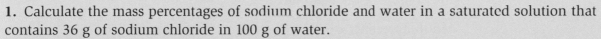

4   Calculate the mass percentages of all substances (solutes and the solvent) in a solution containing 250 mg of potassium chloride, KCl, and 4.75 g of glucose, $C_6H_{12}O_6$, in 45.0 g of water.

In many chemistry problems, volumes of solutions are given instead of masses. To find the mass of a solution from its volume, we need to know the density ($\rho$) of the solution:

$$\rho_{solution} = \frac{m_{solution}}{V_{solution}}$$

and therefore

$$m_{solution} = V_{solution} \times \rho_{solution}$$

The density of pure water varies slightly with temperature. However, these variations are very small (less than 0.003 g cm$^{-3}$ in the range of 0 to 25°C), so in all problems we will assume that $\rho(\text{H}_2\text{O}) = 1.00$ g cm$^{-3}$.

**Question**

5   Calculate the masses of sulfuric acid and water in 0.200 dm³ of an aqueous solution with $\omega(\text{H}_2\text{SO}_4) = 20.0\%$ and $\rho = 1.22$ g cm$^{-3}$.

**Maths skills**

The units of density (usually g cm$^{-3}$ or kg dm$^{-3}$) must match the units of mass (g or kg) and volume (cm³ or dm³). If they do not, we can convert the units as follows:

1 kg = 1000 g

1 g = 1000 mg

1 m³ = 1000 dm³

1 dm³ = 1000 cm³

The use of mass fractions simplifies conversion between masses of dissolved substances and solutions. However, in chemical reactions, substances are consumed or produced according to their amounts (numbers of moles), not masses. In addition, the volumes of solutions are generally easier to measure than their masses. Therefore, we need a concentration expression that relates the amount of a substance to the volume of the solution.

### Molar concentration

*Molar concentration* (*c*), also known as *molarity* or *amount concentration*, is the ratio of the amount of a solute to the volume of the solution:

$$c_{solute} = \frac{n_{solute}}{V_{solution}}$$

The most common units for molar concentration are mol dm⁻³ (which is the same as mol L⁻¹). For very dilute solutions, smaller units (mmol dm⁻³ or μmol dm⁻³) can also be used:

$$1 \text{ mmol dm}^{-3} = 1 \times 10^{-3} \text{ mol dm}^{-3}$$
$$1 \text{ μmol dm}^{-3} = 1 \times 10^{-6} \text{ mol dm}^{-3}$$

The units of molar concentrations are sometimes abbreviated as M (for mol dm⁻³) or mM (for mmol dm⁻³). For example, the expression "2.5 M NaOH" means that each dm³ of the solution contains 2.5 mol of sodium hydroxide.

> **Key term**
>
> **Molar concentration (*c*)** is the ratio of the amount of a solute (in moles) to the volume of the solution (in dm³).

---

### Worked example: Calculating the molar concentration

**2.** Calculate the molar concentration of sodium chloride in its saturated solution, which has a density of 1.20 g cm⁻³.

*Solution*

In the previous worked example, we found that the mass fraction of sodium chloride in its saturated solution is 0.265. Because the concentration is a relative quantity, we can use an arbitrary mass or volume of the solution in our calculations. When we know the mass fraction or mass percentage, it is convenient to take 100 g of the solution. In such case:

$m(NaCl) = 0.265 \times 100 \text{ g} = 26.5 \text{ g}$

$n(NaCl) = 26.5 \text{ g}/58.44 \text{ g mol}^{-1} \approx 0.453 \text{ mol}$

$V(\text{solution}) = 100 \text{ g}/1.20 \text{ g cm}^{-3} \approx 83.3 \text{ cm}^3 = 0.0833 \text{ dm}^3$

$c(NaCl) = 0.453 \text{ mol}/0.0833 \text{ dm}^3 \approx 5.44 \text{ mol dm}^{-3}$.

---

### Question

6   Calculate the molar concentration, in mmol dm⁻³, of bromine in its solution with $\omega(Br_2) = 0.675\%$ and $\rho = 1.01$ g cm⁻³.

A similar approach can be used if we need to calculate the mass percentage of a solute from its molar concentration. In this case it is convenient to take 1 dm³ of the solution (unless a specific volume or mass is already given).

### Dilution and mixing calculations

It is a common practice to store chemicals in the form of concentrated solutions (so-called *stock solutions*) and dilute them to required concentration when needed. Typical problems involving the dilution and mixing of aqueous solutions are discussed on the following pages.

# Worked example: Diluting a solution

**3.** Concentrated hydrochloric acid has a density of 1.18 g cm⁻³ and contains 36.0% by mass of hydrogen chloride. Calculate the volumes of concentrated hydrochloric acid and water (density 1.00 g cm⁻³) that must be mixed together to prepare 500 cm³ of 2.00 mol dm⁻³ hydrochloric acid with $\rho$ = 1.033 g cm⁻³.

*Solution*

The dilution scheme will look as follows:

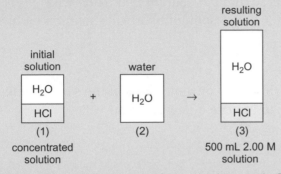

**Figure 2.** Dilution scheme for worked example 3

Although in reality the solute and solvent form a homogeneous mixture, it is helpful to show them separately for the calculation. In figure 2 each solution is represented as two parts: the solute (blue) and the solvent (white). All solutions are numbered, so we can easily refer to them. For example, "the mass of hydrogen chloride in the initial solution" can be written as $m_1(HCl)$.

The dilution scheme shows that $m_1(HCl) = m_3(HCl)$. Indeed, the initial solution is the only source of hydrogen chloride, and all hydrogen chloride from solution 1 becomes a part of solution 3. Similarly, the scheme shows that $m_1(\text{solution}) + m_2(H_2O) = m_3(\text{solution})$. Therefore, we can first calculate the masses of hydrogen chloride and water in solution 3, and then use the equations from the previous section for finding everything else.

$$V_3(\text{solution}) = 500 \text{ cm}^3 = 0.500 \text{ dm}^3$$
$$m_3(\text{solution}) = 500 \text{ cm}^3 \times 1.033 \text{ g cm}^{-3} = 516.5 \text{ g}$$
$$n_3(HCl) = 2.00 \text{ mol dm}^{-3} \times 0.500 \text{ dm}^3 = 1.00 \text{ mol}$$
$$m_3(HCl) = 1.00 \text{ mol} \times 36.46 \text{ g mol}^{-1} = 36.46 \text{ g}.$$

Since $m_1(HCl) = m_3(HCl) = 36.46$ g, we can now find the mass and volume of solution 1:

$$m_1(\text{solution}) = (36.46 \text{ g}/36.0\%) \times 100\% \approx 101.3 \text{ g}$$
$$V_1(\text{solution}) = 101.3 \text{ g}/1.18 \text{ g cm}^{-3} \approx 85.8 \text{ cm}^3.$$

Finally, we can find how much water must be added to solution 1 to prepare solution 3:

$$m_2(H_2O) = m_3(\text{solution}) - m_1(\text{solution}) = 516.5 \text{ g} - 101.3 \text{ g}$$
$$= 415.2 \text{ g}$$
$$V_2(H_2O) = 415.2 \text{ g} \times 1.00 \text{ g cm}^{-3} = 415.2 \text{ cm}^3.$$

Therefore, to prepare 500 cm³ of 2.00 mol dm⁻³ hydrochloric acid, we need to mix together 85.8 cm³ of concentrated hydrochloric acid and 415.2 cm³ of water. Alternatively, we can prepare the same solution by weight, using 101.3 g of concentrated hydrochloric acid and 415.2 g of water.

## Maths skills: Working with concentrations shortcuts

Mass percentage and molar concentration can be quickly converted into one another as follows:

$$\omega\% = \frac{cM}{10\rho}$$

$$c = \frac{10\rho\omega\%}{M}$$

In both cases, the molar concentration ($c$) of the solute must be expressed in mol dm⁻³, the molar mass of the solute ($M$) in g mol⁻¹ and the density ($\rho$) of the solution in g cm⁻³ (or kg dm⁻³).

Use these shortcuts with care, as incorrectly memorized formulae and/or units are the most common source of errors during chemistry examinations.

It is important to note that volumes of solutions are *not* additive. In the example above, 85.8 cm³ + 415.2 cm³ = 501 cm³, but the actual volume of the final solution is 500 cm³. Although this fact is often ignored in chemistry textbooks, you should never add together the volumes of solutions unless they are very dilute or the question explicitly permits addition.

## Question

7   Calculate the volumes of a 25.0% aqueous solution of sodium hydroxide with $\rho = 1.275$ g cm⁻³ and water required to prepare 80.0 cm³ of a 10.0% sodium hydroxide solution with $\rho = 1.109$ g cm⁻³.

Solutions with different concentrations of the same solute can be mixed together to prepare a new solution.

## Worked example: Mixing solutions                                    WE

**4.** Calculate the mass percentage of potassium hydroxide in a solution prepared by mixing together 40 g of a 10% solution of KOH and 20 g of a 40% solution of KOH.

*Solution*

First of all, we need to draw the mixing scheme. For simplicity, here and in all other problems the solutions will be shown as boxes of equal sizes (not to scale):

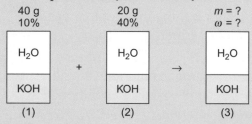

**Figure 3.** Dilution scheme for worked example 4

The scheme shows that $m_1(KOH) + m_2(KOH) = m_3(KOH)$, and $m_1(\text{solution}) + m_2(\text{solution}) = m_3(\text{solution})$. Therefore:

$$m_1(KOH) = (40 \text{ g} \times 10\%)/100\% = 4 \text{ g}$$

$$m_2(KOH) = (20 \text{ g} \times 40\%)/100\% = 8 \text{ g}$$

$$m_3(KOH) = 4 \text{ g} + 8 \text{ g} = 12 \text{ g}$$

$$m_3(\text{solution}) = 40 \text{ g} + 20 \text{ g} = 60 \text{ g}$$

$$\omega_3(KOH) = (12 \text{ g}/60 \text{ g}) \times 100\% = 20\%.$$

 **DP link**

The concepts covered in this section will be useful for **1.3 Reacting masses and volumes** and **8.2 Properties of acids and bases.**

## Question

8   Equal volumes of a 20.0% KOH solution with $\rho = 1.185$ g cm⁻³ and a 50.0% KOH solution with $\rho = 1.515$ g cm⁻³ were mixed together. Calculate the mass percentage of KOH in the final solution.

*Hint: all data in this problem are relative, so you can take an arbitrary volume or mass of either solution.*

## Calculations involving neutralization reactions

Chemical reactions involving acids and bases nearly always take place in solutions. Therefore, we often need to calculate concentrations or volumes of solutions containing these compounds. The simplest example of an acid–base reaction is the neutralization of hydrochloric acid with sodium hydroxide:

$$HCl(aq) + NaOH(aq) \rightarrow NaCl(aq) + H_2O(l)$$

If the amounts of both reactants are the same, the final solution will contain only sodium chloride and water. However, if either of the reactants is in excess, it will not be consumed completely, so some of it will remain in the solution along with the products. For example, if we mix 1.0 dm³ of 0.10 mol dm⁻³ hydrochloric acid and 1.0 dm³ of 0.30 mol dm⁻³ aqueous sodium hydroxide, the resulting solution will consist of water, sodium chloride and unreacted sodium hydroxide:

|            | HCl(aq) | + | NaOH(aq) | → | NaCl(aq) | + | H₂O(l) |
|------------|---------|---|----------|---|----------|---|--------|
| $n_{init}$ | 0.10    |   | 0.30     |   | 0        |   |        |
| $\Delta n$ | −0.10   |   | −0.10    |   | +0.10    |   |        |
| $n_{fin}$  | 0       |   | 0.20     |   | 0.10     |   |        |

Since the initial solutions are very dilute, their densities are approximately 1.0 g cm⁻³ and their volumes are additive. Therefore:

$$V_{fin}(solution) = 1.0 + 1.0 = 2.0 \text{ dm}^3$$
$$c_{fin}(NaCl) = 0.10 \text{ mol}/2.0 \text{ dm}^3 = 0.050 \text{ mol dm}^{-3}$$
$$c_{fin}(NaOH) = 0.20 \text{ mol}/2.0 \text{ dm}^3 = 0.10 \text{ mol dm}^{-3}.$$

### Internal link

The reactions of acids and bases are introduced in **3.1 Classification and nomenclature of inorganic compounds.**

### Question

9  0.20 dm³ of 0.10 mol dm⁻³ sulfuric acid was mixed with an equal volume of 0.10 mol dm⁻³ potassium hydroxide. Calculate the molar concentrations of all solutes in the mixture. Assume that volumes of dilute solutions are additive.

Another common acid–base problem is the calculation of the volume of the solution needed for complete neutralization of an acid or base in the given solution.

### Worked example: Calculating the neutralization volume

**5.** Calculate the volume, in cm³, of 0.50 mol dm⁻³ sodium hydroxide required for complete neutralization of 0.10 dm³ of 0.10 mol dm⁻³ phosphoric acid.

*Solution*

$$H_3PO_4(aq) + 3NaOH(aq) \rightarrow Na_3PO_4(aq) + 3H_2O(l)$$

According to the equation, the amount of sodium hydroxide is three times the amount of phosphoric acid, so:

$$n(H_3PO_4) = 0.10 \text{ dm}^3 \times 0.10 \text{ mol dm}^{-3} = 0.010 \text{ mol}$$
$$n(NaOH) = 3 \times 0.010 \text{ mol} = 0.030 \text{ mol}$$
$$V_{sol}(NaOH) = 0.030 \text{ mol}/0.50 \text{ mol dm}^{-3} = 0.060 \text{ dm}^3 = 60 \text{ cm}^3.$$

**Question**

10 Calculate the volume of 0.25 mol dm⁻³ hydrochloric acid required to neutralize 2.0 dm³ of 0.15 mol dm⁻³ barium hydroxide.

Neutralization reactions are used to determine unknown concentrations of acids and bases by a technique known as *titration* (described in the *Practical skills* box).

## Key term

**Titration** is a method of determining the unknown concentration of a solute (typically an acid or base) by measuring the volumes of reacting solutions.

A **standard solution** is a solution of a reactant with a known concentration.

## Practical skills: Titration

In order to determine the concentration of an acid (or a base), it can be *titrated* against a known solution of a base (or an acid). The solution to be analysed – say an acid – is gradually neutralized by dropwise addition of a solution of a base with known concentration (a so-called *standard solution*). The addition is stopped when the acid is completely consumed. This end-point is usually detected by use of an acid-base indicator that changes colour when the solution becomes neutral.

The mole ratio in which the acid and the base react is known from the equation for the neutralization. If the volumes of both solutions (analysed and standard) are known, the unknown concentration can be determined.

Similarly, the concentration of a base can be determined by neutralizing the base with a standard solution of an acid.

## Worked example: Titration calculations

**6.** Complete neutralization of 10.0 cm³ of a sodium hydroxide solution required 3.75 cm³ of 0.144 mol dm⁻³ sulfuric acid. Determine the concentration, in mol dm⁻³, of sodium hydroxide in the initial solution.

*Solution*

$$2NaOH(aq) + H_2SO_4(aq) \rightarrow Na_2SO_4(aq) + 2H_2O(l)$$

According to the equation, two moles of sodium hydroxide react with one mole of sulfuric acid, so:

$n(H_2SO_4) = 0.00375\ dm^3 \times 0.144\ mol\ dm^{-3} = 0.000540\ mol$

$n(NaOH) = 2 \times 0.000540\ mol = 0.00108\ mol$

$c(NaOH) = 0.00108\ mol/0.0100\ dm^3 = 0.108\ mol\ dm^{-3}$.

## Question

11 Complete neutralization of 20.0 cm³ of a sulfuric acid solution required 9.25 cm³ of 0.120 mol dm⁻³ potassium hydroxide. Calculate the concentration of sulfuric acid in the initial solution.

## DP link

This section is particularly useful preparation for **9.1 Oxidation and reduction** and **D.4 pH regulation of the stomach.**

## 4.3 Solutions of electrolytes

Solutions of many inorganic compounds conduct electricity and can be decomposed by electrolysis. Such compounds are known as *electrolytes* (from Greek *elektron* "amber", which could become electrically charged, + *lytos* "released"). The properties of electrolytes can be explained by their ability to form charged species (cations and anions) that move to opposite electrodes when an electric current is passed through the solution.

**Figure 4.**
Svante Arrhenius
(1859–1927)

In 1884, the Swedish scientist Svante Arrhenius (figure 4) suggested that electrolytes dissociated into ions even in the absence of electric current. According to his *theory of electrolytic dissociation*, dilute aqueous solutions of salts contain only charged species surrounded by molecules of water. Although initially rejected, in the following years this theory was supported by extensive experimental evidence and became universally accepted by the scientific community.

 **Internal link**

Electrolysis is described in **3.3 Redox processes**.

 **Key term**

An **electrolyte** is a compound that dissociates into ions in solution.

The dissociation of electrolytes in aqueous solutions is a result of ion–dipole interactions. In many cases, such interactions are strong enough to break ionic or covalent bonds in the electrolyte and pull the ions apart from one another, as shown in figure 5. Weaker ion–dipole interactions or stronger bonds may lead to partial dissociation and/or low solubility of the electrolyte in water.

 **Internal link**

Ion–dipole interactions are introduced in **2.3 Chemical bonding**.

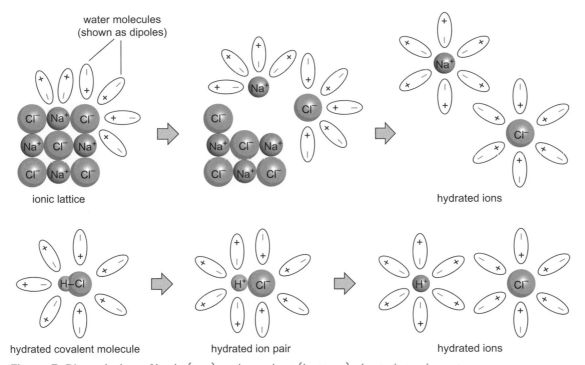

**Figure 5.** Dissociation of ionic (top) and covalent (bottom) electrolytes in water

### Strong and weak electrolytes

*Strong electrolytes* dissociate completely into ions when dissolved in water. Strong electrolytes include all salts, some acids (HCl, HBr, HI, $HNO_3$, $H_2SO_4$, $HClO_3$ and $HClO_4$), all hydroxides of group 1 elements (LiOH, NaOH, KOH, RbOH and CsOH) and most hydroxides of group 2 elements ($Ca(OH)_2$, $Sr(OH)_2$ and $Ba(OH)_2$).

### Worked example: Writing dissociation equations

7. Formulate equations for the dissociation of nitric acid and potassium sulfate in aqueous solution.

*Solution*

$$HNO_3(aq) \rightarrow H^+(aq) + NO_3^-(aq)$$
$$K_2SO_4(aq) \rightarrow 2K^+(aq) + SO_4^{2-}(aq)$$

Note that both electrolytes are strong and thus dissociate irreversibly, which is represented by straight arrows.

If we know the concentration of a strong electrolyte in solution, we can easily find the concentrations of all ions in that solution.

### Worked example: Calculating concentrations of ions

8. Determine the concentrations of all the ions in a 0.10 mol dm$^{-3}$ solution of iron(III) sulfate:

*Solution*

$$Fe_2(SO_4)_3(aq) \rightarrow 2Fe^{3+}(aq) + 3SO_4^{2-}(aq)$$

Each formula unit of iron(III) sulfate produces two iron(III) ions and three sulfate ions, so:

$$c(Fe^{3+}(aq)) = 2 \times 0.10 = 0.20 \text{ mol dm}^{-3}$$
$$c(SO_4^{2-}(aq)) = 3 \times 0.10 = 0.30 \text{ mol dm}^{-3}.$$

The total concentration of ions in this solution will be 0.20 + 0.30 = 0.50 mol dm$^{-3}$.

Note that the solution does not contain any $Fe_2(SO_4)_3$ molecules, as iron(III) sulfate is a salt and thus a strong electrolyte.

### Question

12 Write out the dissociation scheme and calculate the total concentration of ions in a 0.15 mol dm$^{-3}$ solution of ammonium phosphate.

## Key term

A **strong electrolyte** dissociates completely into ions when it is dissolved in water.

A **weak electrolyte** dissociates reversibly in aqueous solutions, so only a fraction of the electrolyte is present in solution as ions.

The **degree of ionization** ($\alpha$) is the ratio of the concentration of ionized weak electrolyte ($c_{ion}$) to the total concentration ($c$) of electrolyte in the solution; $\alpha = \frac{c_{ion}}{c} \times 100\%$.

Nearly all acids and bases not listed above are *weak electrolytes*. Their dissociation in aqueous solutions is a reversible process; for example:

$$HF(aq) \rightleftharpoons H^+(aq) + F^-(aq)$$

The equilibrium sign $\rightleftharpoons$ in this equation means that all three species, HF(aq), H$^+$(aq) and F$^-$(aq), are present in the solution at the same time. Typically, only a small proportion of a weak electrolyte exists in the form of ions. This proportion is known as the *degree of ionization* ($\alpha$), which can be defined as the ratio of the concentration of ionized electrolyte ($c_{ion}$) to the total electrolyte concentration ($c$) in the solution:

$$\alpha = \frac{c_{ion}}{c} \times 100\%$$

### Worked example: Calculating the degree of ionization

**9.** According to experimental data, a 1.0 mol dm$^{-3}$ solution of hydrogen fluoride contains 0.026 mol dm$^{-3}$ of hydrogen ions. Determine the degree of ionization of hydrogen fluoride in this solution.

**Solution**

Each molecule of ionized hydrogen fluoride produces one H$^+$(aq) ion, so:

$$\alpha(\text{HF}) = \frac{c(\text{H}^+(\text{aq}))}{c(\text{HF})} \times 100\% = \frac{0.026 \text{ mol dm}^{-3}}{1.0 \text{ mol dm}^{-3}} \times 100\% = 2.6\%$$

### Question

13 Ammonia is a weak base that undergoes ionization as follows:

$$\text{NH}_3(\text{aq}) + \text{H}_2\text{O}(\text{l}) \rightleftharpoons \text{NH}_4^+(\text{aq}) + \text{OH}^-(\text{aq})$$

Determine the concentration of hydroxide ions in a 0.050 mol dm$^{-3}$ solution of ammonia if its degree of ionization in this solution is 3.0%.

The degree of ionization of strong electrolytes is assumed to be 100%. For weak electrolytes, the degree of ionization increases with dilution: the lower the concentration of the electrolyte, the greater its degree of ionization. However, even very dilute solutions of weak electrolytes still contain some undissociated molecules, while strong electrolytes remain completely ionized at any concentrations.

Weak acids with more than one exchangeable hydrogen undergo stepwise dissociation:

$$\text{H}_3\text{PO}_4(\text{aq}) \rightleftharpoons \text{H}^+(\text{aq}) + \text{H}_2\text{PO}_4^-(\text{aq})$$
$$\text{H}_2\text{PO}_4^-(\text{aq}) \rightleftharpoons \text{H}^+(\text{aq}) + \text{HPO}_4^{2-}(\text{aq})$$
$$\text{HPO}_4^{2-}(\text{aq}) \rightleftharpoons \text{H}^+(\text{aq}) + \text{PO}_4^{3-}(\text{aq})$$

However, only the first step produces any significant concentration of hydrogen ions, so the second and further steps can usually be ignored.

### Question

14 Write out the scheme of stepwise dissociation for carbonic acid (refer to table 1 in *3 Inorganic chemistry*).

Nearly all weak inorganic bases are insoluble in water, so they cannot produce hydroxide ions in aqueous solutions. The only exception is ammonia, which is readily soluble in water and undergoes ionization as shown in question 13.

Water itself is a very weak electrolyte that dissociates into hydrogen cations and hydroxide anions:

$$\text{H}_2\text{O}(\text{l}) \rightleftharpoons \text{H}^+(\text{aq}) + \text{OH}^-(\text{aq})$$

At room temperature, very few molecules of water (1 in 2,000,000,000) exist in the form of ions. Nevertheless, the dissociation of water is a very important process (see *6.3 The concept of pH*).

In contrast to electrolytes, *non-electrolytes* do not produce any ions, either in solutions or in their individual form. Many inorganic non-electrolytes are elementary substances and oxides, such as molecular oxygen and carbon monoxide. Organic non-electrolytes include hydrocarbons, alcohols and sugars.

**Internal link**

The organic non-electrolytes named here are described in detail in **7 Organic chemistry**.

**Practical skills: Determining electrolytic strength**

The strength of an electrolyte can be determined experimentally by measuring the electrical conductivity of its solution. In a typical experiment, the solution is placed into to a beaker with two electrodes that form a series circuit with a battery and a light bulb (figure 6). The electric current in the circuit is proportional to the number of ions in the solution. A strong electrolyte will produce more ions than a weak electrolyte of the same concentration, so the bulb will glow brighter in the first case and dimmer in the second case. A very weak electrolyte, such as water, will produce almost no ions, so the bulb will not light up at all. Similarly, the bulb will remain off when the beaker is filled with an aqueous solution of a non-electrolyte, such as sugar.

**Figure 6.** The electrical conductivity test

**Internal link**

The strengths of acids and bases can also be determined by measuring the thermal effects of their neutralization reactions, which are discussed in **5.1 Thermochemistry**.

### Ionic equations

Chemical reactions in solutions of electrolytes often involve ions rather than neutral molecules. Such reactions are best described by *ionic equations*.

**Worked example: Ionic equations**

**10.** Describe the neutralization of hydrogen chloride with sodium hydroxide using the molecular, total ionic and net ionic equations.

*Solution*

The molecular equation is as follows:

$$HCl(aq) + NaOH(aq) \rightarrow NaCl(aq) + H_2O(l)$$

Three of the four substances in this equation (HCl, NaOH and NaCl) are strong electrolytes, which will be completely ionized in the solution:

$$HCl(aq) \rightarrow H^+(aq) + Cl^-(aq)$$
$$NaOH(aq) \rightarrow Na^+(aq) + OH^-(aq)$$
$$NaCl(aq) \rightarrow Na^+(aq) + Cl^-(aq)$$

In contrast, water is a very weak electrolyte, so almost all its molecules will remain undissociated. Therefore, we can replace the molecular formulae of hydrogen chloride, sodium hydroxide and sodium chloride with ions while leaving water in its molecular form:

$$H^+(aq) + Cl^-(aq) + Na^+(aq) + OH^-(aq) \rightarrow Na^+(aq) + Cl^-(aq) + H_2O(l)$$

The resulting *total ionic equation* includes all species (ions and molecules) that physically exist in the solution. However, the ions $Cl^-(aq)$ and $Na^+(aq)$ are present on both sides of the equation. These *spectator ions* do not participate in any chemical changes and thus can be cancelled out:

$$H^+(aq) + Cl^-(aq) + Na^+(aq) + OH^-(aq) \rightarrow Na^+(aq) + Cl^-(aq) + H_2O(l)$$

The remaining species produce the *net ionic equation*:

$$H^+(aq) + OH^-(aq) \rightarrow H_2O(l)$$

---

**DP ready** **Approaches to learning**

You may find it useful to formulate a strategy when solving typical problems. Here is an example of such a strategy for constructing net ionic equations.

1. Write and balance the molecular equation, showing the states of all reactants and products.

2. Using the solubility table in the appendix, identify all strong electrolytes that are soluble in water and replace their molecular formulae with ions.

3. Leave all weak electrolytes, non-electrolytes, gases and insoluble substances in molecular form.

4. Identify and cancel out the spectator ions.

5. Check that the net ionic equation is balanced for both elements and charges.

> **Key term**
>
> In **ionic equations** strong electrolytes are shown as their ions.
>
> A **total ionic equation** includes all species (ions and molecules) that physically exist in a solution during a reaction.
>
> A **net ionic equation** includes only the ions participating in chemical changes.

**Question**

15 Write the molecular, total ionic and net ionic equations for the reaction between nitric acid and potassium hydroxide.

> **Key term**
>
> **Spectator ions** are ions that are present in a solution of reactants but that do not participate in the reaction.

Net ionic equations are very important in chemistry, as they contain only those species that are essential for the reaction. In particular, the last worked example shows that a reaction between a strong acid and a strong base involves only $H^+(aq)$ and $OH^-(aq)$ ions, and produces a weak electrolyte (water). This makes the reaction almost irreversible, as the opposite process (dissociation of water) occurs only to a very small extent.

Other irreversible processes may include the formation of insoluble substances (*precipitation reactions*) and gases (*effervescent reactions*). In both cases, the driving force for the process is the removal of products from the reaction mixture. For example, when aqueous solutions of silver nitrate and sodium chloride are mixed together, a white precipitate of silver chloride forms:

$$AgNO_3(aq) + NaCl(aq) \rightarrow NaNO_3(aq) + AgCl(s)$$

Silver chloride is insoluble in water (see the solubility table in the appendix). Therefore, it leaves the solution and makes the reverse

> **Key term**
>
>
>
> In a **precipitation reaction**, an insoluble compound is formed, so a solid precipitates out of solution.
>
> In an **effervescent reaction**, a gas is formed and bubbles out of solution.

reaction unfavourable. This process can also be represented by ionic equations:

$$Ag^+(aq) + NO_3^-(aq) + Na^+(aq) + Cl^-(aq) \rightarrow Na^+(aq) + NO_3^-(aq) + AgCl(s)$$

$$Ag^+(aq) + Cl^-(aq) \rightarrow AgCl(s)$$

Note that insoluble substances are represented by molecular formulae, as they do not produce significant amounts of ions in the solution.

Effervescent reactions of acids with metal carbonates or hydrogencarbonates are often used for detecting the presence of these substances in solutions. Similar reactions take place when acids are mixed with metal sulfides:

$$2HCl(aq) + Na_2S(aq) \rightarrow 2NaCl(aq) + H_2S(g)$$

$$2H^+(aq) + 2Cl^-(aq) + 2Na^+(aq) + S^{2-}(aq) \rightarrow 2Na^+(aq) + 2Cl^-(aq) + H_2S(g)$$

$$2H^+(aq) + S^{2-}(aq) \rightarrow H_2S(g)$$

Hydrogen sulfide, a toxic gas with the distinctive smell of rotten eggs, escapes from the reaction mixture and makes the process irreversible.

> **Internal link**
>
> For the reactions of acids with metal carbonates and hydrogencarbonates, which give off carbon dioxide, see 6.2 Classification and properties of acids and bases.

> **Question** Q
>
> 16 Write the molecular, total ionic and net ionic equations for the following reactions: a) barium chloride with potassium sulfate; b) hydrogen chloride with sodium hydrogensulfide.

> **Internal link**
>
> Redox reactions are described in 3.3 Redox processes.

Ionic equations are particularly useful for describing redox processes. The oxidation and reduction half-equations can be balanced independently and then combined together to produce the net ionic equation. After that, the molecular equation can be deduced by adding spectator ions and combining some pairs of ions into molecular species. This approach, known as the *half-equation method*, requires some practice but usually makes the balancing very straightforward.

## Worked example: The half-equation method for balancing redox equations   WE

**11.** The reaction between potassium dichromate and potassium iodide in the presence of sulfuric acid was given in worked example 4 in the previous chapter. Deduce the balanced equation for this reaction using the half-equation method.

*Solution*

1. Unbalanced molecular equation (the states are omitted for clarity):

$$K_2Cr_2O_7 + KI + H_2SO_4 \rightarrow K_2SO_4 + Cr_2(SO_4)_3 + I_2 + H_2O$$

2. Unbalanced total ionic equation (all stoichiometric coefficients removed):

$$K^+ + Cr_2O_7^{2-} + I^- + H^+ + SO_4^{2-} \rightarrow K^+ + SO_4^{2-} + Cr^{3+} + I_2 + H_2O$$

3. Unbalanced net ionic equation (spectator ions are cancelled out):

$$Cr_2O_7^{2-} + I^- + H^+ \rightarrow Cr^{3+} + I_2 + H_2O$$

4. Oxidation half-equation (note that ionic charges are used instead of oxidation numbers):

$$2I^- \rightarrow I_2 + 2e^-$$

5. Reduction half-equation (note that actual species are used instead of individual elements):

$$Cr_2O_7^{2-} + 14H^+ + 6e^- \rightarrow 2Cr^{3+} + 7H_2O$$

**6.** Electron balance:

$$2I^- \rightarrow I_2 + 2e^- \qquad\qquad\qquad \times\ 3$$
$$Cr_2O_7^{2-} + 14H^+ + 6e^- \rightarrow 2Cr^{3+} + 7H_2O \qquad \times\ 1$$

**7.** Balanced net ionic equation:

$$Cr_2O_7^{2-} + 6I^- + 14H^+ \rightarrow 2Cr^{3+} + 3I_2 + 7H_2O$$

**8.** Balanced molecular equation (note that $14H^+$ becomes $7H_2SO_4$ and $2Cr^{3+}$ becomes $Cr_2(SO_4)_3$ while all other stoichiometric coefficients remain unchanged):

$$K_2Cr_2O_7 + 6KI + 7H_2SO_4 \rightarrow 4K_2SO_4 + Cr_2(SO_4)_3 + 3I_2 + 7H_2O$$

## Question

17 Deduce the balanced equation for the following redox reaction using the half-equation method.

$$KMnO_4 + KNO_2 + H_2SO_4 \rightarrow K_2SO_4 + MnSO_4 + KNO_3 + H_2O$$

Half-equations reflect the nature of chemical changes in solutions of electrolytes and emphasize the central role of ionic species in these processes. We will continue the discussion of ionic equations in *6 Acids and bases*, where you will learn more about the properties of acids and bases, and their behaviour in aqueous solutions.

## Chapter summary

In this chapter, you have learned about solutions, concentration expressions, electrolytes and ionic equations. Make sure that you have a working knowledge of the following concepts and definitions:

☐ Solutions are homogeneous mixtures of two or more components.

☐ Solubility is the maximum mass, volume or amount of the solute that can be dissolved in a unit mass or volume of the solvent.

☐ Solubilities of ionic solids are not affected by pressure but increase with temperature.

☐ Solubilities of gases increase with pressure and decrease with temperature.

☐ An unsaturated solution can dissolve an additional quantity of the solute while a saturated solution cannot dissolve any more solute.

☐ The proportion of the solute is low in a dilute solution and high in a concentrated solution.

☐ Saturation and concentration are not related to one another, so a saturated solution could be very dilute while an unsaturated solution could be highly concentrated.

☐ The mass percentage ($\omega\%$) of a solute is its mass fraction ($\omega$) expressed as a percentage.

☐ The molar concentration ($c$) of a solute is the ratio of its amount to the volume of the solution.

☐ The units of molar concentration are $mol\ dm^{-3}$ (sometimes abbreviated as "M").

☐ Unknown concentrations of acids and bases can be determined by titration.

☐ Strong electrolytes are fully ionized in aqueous solutions, while weak electrolytes undergo only partial ionization.

☐ The degree of ionization ($\alpha$) of an electrolyte is the ratio of the concentration of the ionized electrolyte to its total concentration.

☐ The degree of ionization of weak electrolytes increases with dilution.

- In ionic equations, soluble strong electrolytes are shown as ions while all other species are shown as molecules.
- Spectator ions do not participate in chemical changes and thus do not appear in net ionic equations.
- Chemical reactions producing precipitates, gases, water or other weak electrolytes are usually irreversible.
- Redox equations can be balanced using the half-equation method.

## Additional problems

1. A saturated solution of ammonium nitrate contains more salt than water (table 1). Explain why it is more convenient to choose water as the solvent.

2. Using table 1, calculate the mass percentage and molar concentration of potassium bromide in its saturated solution with a density of $1.37 \text{ g cm}^{-3}$.

3. A sudden drop in atmospheric pressure can cause decompression sickness, as explained at the end of *4.1 Composition and classification of solutions*. Suggest why this health condition is associated with nitrogen but not with oxygen, despite the fact that both gases have similar solubilities in body tissues.

4. Normal saline contains 9.0 g of sodium chloride in $1.0 \text{ dm}^3$ of the solution and has a density of $1.005 \text{ g cm}^{-3}$. Calculate the mass percentage and the molar concentration of sodium chloride in this solution.

5. Nitric acid is often sold as a $15 \text{ mol dm}^{-3}$ solution in water with a density of $1.41 \text{ g cm}^{-3}$. Calculate the mass percentage of nitric acid in this solution.

6. Calculate the volumes of water and the nitric acid solution from the previous problem that need to be mixed together to prepare 250 g of a 10% solution of nitric acid.

7. Calculate the molar concentrations of all solutes in a solution prepared by mixing together $0.40 \text{ dm}^3$ of $0.25 \text{ mol dm}^{-3}$ hydrochloric acid and $0.10 \text{ dm}^3$ of $0.60 \text{ mol dm}^{-3}$ potassium hydroxide. Assume that volumes of dilute solutions are additive.

8. The complete neutralization of $10.0 \text{ cm}^3$ of a hydrogen bromide solution required $12.4 \text{ cm}^3$ of a $0.482 \text{ mol dm}^{-3}$ sodium hydroxide solution. Calculate the concentration of hydrogen bromide in the initial solution.

9. Describe what happens when sodium chloride is dissolved in water. Discuss whether this process is a physical or chemical change.

10. Write out the dissociation schemes for the following electrolytes in aqueous solutions:
a) $(NH_4)_2SO_4$; b) $Al(NO_3)_3$; c) $HNO_2$; d) $H_3PO_4$.

11. Sulfurous acid is unstable and readily decomposes into sulfur(IV) oxide and water. Write the molecular, total ionic and net ionic equations for the reactions between: a) solid magnesium sulfite and aqueous sulfuric acid; b) aqueous sodium hydrogensulfite and aqueous hydrogen chloride.

12. Suggest whether the following compounds will react with hydrogen sulfide: a) sodium nitrate; b) silver nitrate. Explain your answer by writing molecular, total ionic and net ionic equations.

13. Write the molecular, total ionic and net ionic equations for all possible reactions between aqueous solutions of potassium carbonate, sodium chloride and silver nitrate. Explain why two of these compounds do not react with each other.

14. Deduce the balanced equations for the following redox reactions using the half-equation method.
a) $KMnO_4(aq) + HCl(aq) \rightarrow KCl(aq) + MnCl_2(aq) + Cl_2(g) + H_2O(l)$
b) $FeCl_2(aq) + H_2O_2(aq) + HCl(aq) \rightarrow FeCl_3(aq) + H_2O(l)$
Note that hydrogen peroxide ($H_2O_2$) is a weak electrolyte.

> *A theory is the more impressive the greater the simplicity of its premises, the more different kinds of things it relates, and the more extended its area of applicability. Therefore the deep impression that classical thermodynamics made upon me. It is the only physical theory of universal content which I am convinced will never be overthrown, within the framework of applicability of its basic concepts.*
>
> **Albert Einstein, *Autobiographical Notes* (1949)**

## Chapter context

In this chapter, we will discuss the most basic laws that govern all **chemical changes**. There are three sections: **thermochemistry**, **chemical kinetics** and **chemical equilibrium**.

## Learning objectives

In this chapter you will learn about:

→ the **heat energy** associated with chemical reactions

→ the ways energy is **measured** or **calculated**.

→ **rates** of chemical reactions

→ **collision theory** and **catalysis**

→ **reversible reactions** and factors affecting their direction and yield.

 **Key terms introduced**

→ Exothermic and endothermic reactions

→ The system and the surroundings

→ Enthalpy ($H$)

→ Standard enthalpy changes; standard enthalpies of formation ($\triangle H_f^\circ$) and combustion ($\triangle H_c^\circ$)

→ Bond dissociation energy (bond enthalpy, BE)

→ Entropy ($S$)

→ Average ($v_{avr}$) and instantaneous ($v_{inst}$) reaction rates

→ Phases

→ Catalysts, intermediates and activation energy ($E_a$)

→ Homogeneous and heterogeneous catalysis

→ Chemical equilibrium

→ Equilibrium constant ($K_c$) and reaction quotient ($Q_c$)

→ Le Châtelier's principle

## 5.1 Thermochemistry

As you already know, chemical changes can release or absorb energy in the form of heat. Exothermic processes transfer thermal energy from the reaction mixture (the *system*) to the external environment (the *surroundings*), while endothermic processes transfer thermal energy in the opposite direction, from the surroundings to the system.

 **DP link**

5 Energetics/thermochemistry

 **Internal link**

Exothermic and endothermic reactions were introduced in **3.2 Classification of chemical reactions.**

| **DP ready** | **Nature of science** |

*Chemical thermodynamics* is an interdisciplinary science that studies the energy changes associated with chemical reactions and certain physical processes, such as the change of state of reacting species. A branch of chemical thermodynamics, *thermochemistry*, deals with thermal energy released or consumed in chemical transformations. A broader term, *chemical energetics*, includes thermochemistry and certain aspects of chemical kinetics (see also *5.2 Chemical kinetics*).

**Key term**

**Exothermic** reactions produce heat.

**Endothermic** reactions consume heat.

The **system** is the reaction mixture, and the **surroundings** are the rest of the world, including the container of the mixture.

## Enthalpy changes

Although the absolute heat content of a system, its *enthalpy* (*H*), cannot be determined, the change in enthalpy (Δ*H*) in any particular process is measurable: in very simple terms it is the amount of energy going into the system minus the amount of energy leaving. More precisely, Δ*H* is equal to the amount of work (*W*) done on the system minus the amount of heat (*Q*) lost by the system to the surroundings:

$$\Delta H = W - Q$$

At constant pressure, *W* = 0, and so Δ*H* = −*Q*; therefore, for an exothermic reaction in a *closed system* *Q* > 0 and Δ*H* < 0 while for an endothermic reaction in a closed system *Q* < 0 and Δ*H* > 0 (figure 1).

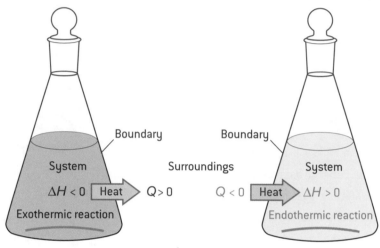

**Figure 1.** Thermal effects and enthalpy changes in a closed system

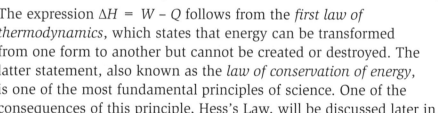

**DP ready | Nature of science**

**The first law of thermodynamics**

The expression Δ*H* = *W* − *Q* follows from the *first law of thermodynamics*, which states that energy can be transformed from one form to another but cannot be created or destroyed. The latter statement, also known as the *law of conservation of energy*, is one of the most fundamental principles of science. One of the consequences of this principle, Hess's Law, will be discussed later in this topic.

Any chemical change that takes place under defined conditions and involves defined amounts of reactants will have a specific Δ*H* value and thus produce or consume a specific amount of thermal energy. For example, the combustion of 1 mol of hydrogen at SATP (298 K and 100 kPa) will release 285.8 kJ of heat. This process can be represented by the following *thermochemical equation*:

$$H_2(g) + 0.5O_2(g) \rightarrow H_2O(l) + 285.8 \text{ kJ}$$

In this equation, the thermal effect is shown as heat (*Q* = +285.8 kJ) released to the environment. Alternatively, the same effect can be shown as the *standard enthalpy change* of the reaction ($\Delta H_{298}^{\circ}$):

$$H_2(g) + 0.5O_2(g) \rightarrow H_2O(l) \quad \Delta H_{298}^{\circ} = -285.8 \text{ kJ}$$

The superscript symbol ° in $\Delta H_{298}^{\circ}$ refers to the standard pressure (100 kPa) and standard states of all participating species (gaseous for hydrogen and oxygen, liquid for water).

Notice that in the notation used above for standard enthalpy change $\Delta H^{\circ}_{298}$, the temperature (298 K) is shown as a subscript index. However, in the IB Chemistry syllabus this index is omitted, and the temperature is assumed to be 298 K unless stated otherwise. From now on, we will follow this convention and represent standard enthalpy changes at $T = 298$ K as $\Delta H^{\circ}$ instead of $\Delta H^{\circ}_{298}$.

When writing thermochemical equations, you must always show the states of all reactants and products under the specified conditions. If the combustion of hydrogen produced steam instead of liquid water, the enthalpy change of the process would be significantly lower (−241.8 kJ instead of −285.8 kJ), as some energy would be required to prevent water vapour from condensing into liquid:

$$H_2(g) + 0.5O_2(g) \rightarrow H_2O(g) \quad \Delta H^{\circ} = -241.8 \text{ kJ}$$

The stoichiometric coefficients in thermochemical equations must also correspond to specified thermal effects. For example, if we double all coefficients in the previous equation, the thermal effect will also double:

$$2H_2(g) + O_2(g) \rightarrow 2H_2O(g) \quad \Delta H^{\circ} = -483.6 \text{ kJ}$$

## Hess's Law

In 1840, the Swiss-born Russian chemist Germain Hess discovered that thermal effects of chemical reactions depended only on the initial and final states of the system but not on the reaction pathway. This observation, now known as *Hess's Law*, allows the use of thermochemical equations with known thermal effects for calculating enthalpy changes of other processes.

To illustrate Hess's Law, we will find the standard enthalpy of vaporization of water, which can be represented by the following equation:

$$H_2O(l) \rightarrow H_2O(g) \qquad \Delta H^{\circ} = x \text{ kJ} \qquad (1)$$

We already know that the formation of liquid water from gaseous hydrogen and oxygen produces more heat than the formation of water vapour from the same reactants:

$$H_2(g) + 0.5O_2(g) \rightarrow H_2O(l) \quad \Delta H^{\circ} = -285.8 \text{ kJ} \qquad (2)$$
$$H_2(g) + 0.5O_2(g) \rightarrow H_2O(g) \quad \Delta H^{\circ} = -241.8 \text{ kJ} \qquad (3)$$

Therefore, we can expect that the vaporization of water will be an endothermic process ($x > 0$), so the enthalpy of water vapour will be higher than that of liquid water (figure 3):

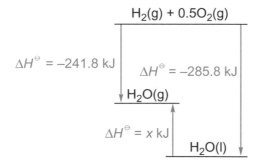

Figure 3. Enthalpy diagram for vaporization of water

**Maths skills: Units of standard enthalpy changes**

When a standard enthalpy change refers to an individual substance, it is assumed that 1 mol of the substance is involved, and so the $\Delta H^{\circ}$ value is usually expressed in kJ mol$^{-1}$. For example, we say that the standard enthalpy of combustion of hydrogen gas is −285.8 kJ mol$^{-1}$, as we are talking about only one species, $H_2(g)$. However, when we represent the same process by a chemical equation, $H_2(g) + 0.5O_2(g) \rightarrow H_2O(l)$, the $\Delta H^{\circ}$ value must be expressed in kJ instead of kJ mol$^{-1}$, as we are talking about more than one mole of participating species.

As follows from figure 3, water vapour can be produced from gaseous hydrogen and oxygen either directly ($\Delta H^\circ$ = –241.8 kJ) or in two steps, where liquid water is initially formed ($\Delta H^\circ$ = –285.8 kJ) and then turned into vapour ($\Delta H^\circ$ = $x$ kJ). Since the enthalpy change is independent of the reaction pathway,

$$-241.8 = -285.8 + x$$
$$\text{so } x = 44.0.$$

Therefore, the standard enthalpy of vaporization of water at 298 K is +44.0 kJ mol$^{-1}$.

The same value for the standard enthalpy of vaporization of water can be obtained by transforming equations (2) and (3) into equation (1).

To do so, we need to look at the reactants and products in each equation. The reactant in equation (1), liquid water, is the product in equation (2), so equation (2) must be reversed. This will change the sign of its $\Delta H^\circ$ value:

$$H_2O(l) \rightarrow H_2(g) + 0.5O_2(g) \quad \Delta H^\circ = +285.8 \text{ kJ} \tag{2'}$$

The product in equation (1), water vapour, is also the product in equation (3), so equation (3) can be used as written:

$$H_2(g) + 0.5O_2(g) \rightarrow H_2O(g) \quad \Delta H^\circ = -241.8 \text{ kJ} \tag{3}$$

Now we can add equations (2') and (3) together:

$$H_2O(l) + \cancel{H_2(g)} + \cancel{0.5O_2(g)} \rightarrow \cancel{H_2(g)} + \cancel{0.5O_2(g)} + H_2O(g)$$

$$\Delta H^\circ = +285.8 + (-241.8) \text{ kJ}$$

$$H_2O(l) \rightarrow H_2O(g) \quad \Delta H^\circ = +44.0 \text{ kJ mol}^{-1} \tag{1}$$

You can use any of the above methods for solving thermochemistry problems. However, it is good practice to draw the enthalpy diagram first, calculate the $\Delta H^\circ$ value as shown in figure 3, and then check your answer by combining thermochemical equations with known standard enthalpy changes.

## Question

1  Sulfur(VI) oxide can be synthesized from elementary sulfur in two steps:

$$S(s) + O_2(g) \rightarrow SO_2(g) \quad \Delta H^\circ = -296.8 \text{ kJ}$$
$$SO_2(g) + 0.5O_2(g) \rightarrow SO_3(g) \quad \Delta H^\circ = -98.9 \text{ kJ}$$

Calculate the standard enthalpy change for the following process:

$$SO_3(g) \rightarrow S(s) + 1.5O_2(g)$$

## Key term

**Standard enthalpy of formation** ($\Delta H_f^\circ$) is the standard enthalpy change for the formation of 1 mol of a substance from its constituent elements, with all participating species in their standard states.

## Calculations involving standard enthalpy of formation

Thermochemical calculations can be simplified by using enthalpy values associated with individual substances. The most common type of enthalpy value used for this purpose is the *standard enthalpy of formation* ($\Delta H_f^\circ$), which is the standard enthalpy change for the formation of 1 mol of a substance from its constituent elements, with all participating species in their standard states.

For example, the standard enthalpy of formation of liquid water (represented by equation (2) on page 113) has a value of −285.8 kJ mol⁻¹.

Selected $\Delta H_f^\circ$ values are given in table 1. Note that the standard enthalpies of formation for the most stable allotropes of elementary substances are assumed to be 0 kJ mol⁻¹, as such substances cannot be formed from more stable species.

**Internal link**

Allotropes are discussed in **2.3 Chemical bonding**.

**Table 1.** Standard enthalpies of formation of selected substances at 298 K

| Substance | $\Delta H_f^\circ$ / kJ mol⁻¹ | Substance | $\Delta H_f^\circ$ / kJ mol⁻¹ |
|---|---|---|---|
| C(s, graphite) | 0 | $H_2SO_4$(l) | −814.0 |
| C(s, diamond) | +1.9 | NaOH(s) | −425.8 |
| $CH_4$(g) | −74.0 | $NH_3$(g) | −45.9 |
| CO(g) | −110.5 | $NH_4Cl$(s) | −314.4 |
| $CO_2$(g) | −393.5 | NO(g) | +91.3 |
| $Ca(OH)_2$(s) | −985.2 | $NO_2$(g) | +33.2 |
| $H_2$(g) | 0 | $N_2O_4$(g) | +11.1 |
| HCl(g) | −92.3 | $O_2$(g) | 0 |
| $HNO_3$(l) | −174.1 | $O_3$(g) | +142.7 |
| $H_2O$(l) | −285.8 | $SO_2$(g) | −296.8 |
| $H_2O$(g) | −241.8 | $SO_3$(g) | −395.7 |

Any equation that represents a standard enthalpy of formation of a particular substance must:

- be balanced
- have the substance in question as the only product
- produce 1 mol of the substance (have a stoichiometric coefficient of 1)
- start with all reactants as elements in their standard states (typically at 100 kPa and 298 K)
- refer to the most stable allotrope if an element forms several allotropes.

These rules are illustrated by the following examples:

$0.5N_2$(g) + $1.5H_2$(g) → $NH_3$(g)        $\Delta H_f^\circ(NH_3(g))$ = −45.9 kJ mol⁻¹

C(s, graphite) + $0.5O_2$(g) → CO(g)        $\Delta H_f^\circ(CO(g))$ = −110.5 kJ mol⁻¹

Na(s) + $0.5O_2$(g) + $0.5H_2$(g) → NaOH(s)   $\Delta H_f^\circ(NaOH(s))$ = −425.8 kJ mol⁻¹

$1.5O_2$(g) → $O_3$(g)        $\Delta H_f^\circ(O_3(g))$ = +142.7 kJ mol⁻¹

**Question**

2  State the equations that represent $\Delta H_f^\circ$ for the following substances: **a)** C(s, diamond); **b)** HCl(g); **c)** $HNO_3$(l); **d)** $Ca(OH)_2$(s).

Standard enthalpies of formation can be combined together using Hess's Law in the same way as standard enthalpy changes of thermochemical equations. If we know $\Delta H_f^\circ$ values for all reactants and products, the enthalpy change for the whole reaction can be found as follows (figure 4):

$$\Delta H^\circ(\text{reaction}) = \Sigma\Delta H_f^\circ(\text{products}) - \Sigma\Delta H_f^\circ(\text{reactants})$$

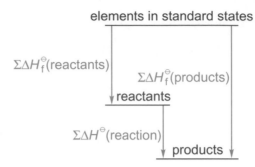

**Figure 4.** Calculating an enthalpy change from standard enthalpies of formation

## Worked example: Calculating the enthalpy change from standard enthalpies of formation

**1.** Determine the $\Delta H^\circ$ value for the following process:

$$4NO_2(g) + 2H_2O(l) + O_2(g) \rightarrow 4HNO_3(l)$$

*Solution*

According to table 1, standard enthalpies of formation of $NO_2(g)$, $H_2O(l)$ and $HNO_3(l)$ are $+33.2$, $-285.8$ and $-174.1$ kJ mol$^{-1}$, respectively. Since $O_2(g)$ is the most stable allotrope of elementary oxygen, its standard enthalpy of formation is 0 kJ mol$^{-1}$. Therefore:

$$\Sigma\Delta H_f^\circ(\text{products}) = 4 \times (-174.1) = -696.4 \text{ kJ}$$

$$\Sigma\Delta H_f^\circ(\text{reactants}) = 4 \times 33.2 + 2 \times (-285.8) + 0$$
$$= -438.8 \text{ kJ}$$

$$\Delta H^\circ(\text{reaction}) = \Sigma\Delta H_f^\circ(\text{products}) - \Sigma\Delta H_f^\circ(\text{reactants})$$
$$= -696.4 - (-438.8) = -257.6 \text{ kJ}.$$

Note that the units have changed from kJ mol$^{-1}$ to kJ, as the total amount of substance produced in the reaction is not 1 mol.

## Question

3   Using the data from table 1, calculate standard enthalpy changes for the following reactions:
a) $NH_3(g) + HCl(g) \rightarrow NH_4Cl(s)$; b) $2CO(g) + 2NO(g) \rightarrow 2CO_2(g) + N_2(g)$.

## Key term

**Standard enthalpy of combustion** $(\Delta H_c^\circ)$ is the enthalpy change for the reaction in which 1 mol of the substance is completely burnt in excess oxygen under standard conditions.

### Calculations involving standard enthalpy of combustion

Combustible substances are often characterized by the *standard enthalpy of combustion*, $\Delta H_c^\circ$, which is the enthalpy change for the reaction where 1 mol of the substance is completely burnt in excess oxygen under standard conditions. The combustion products can include water, carbon dioxide, molecular nitrogen and other substances, as shown in table 2. If atoms of oxygen are present in the original compound, they do not form any additional products but simply reduce the amount of oxygen gas required for the combustion.

**Table 2.** Typical combustion products of elements and combustion reactions of compounds

| Element | Product | Example |
|---|---|---|
| C | $CO_2(g)$ | $CH_4(g) + 2O_2(g) \rightarrow CO_2(g) + 2H_2O(l)$ |
| H | $H_2O(l)$ | |
| N | $N_2(g)$ | $NH_3(g) + 0.75O_2(g) \rightarrow 0.5N_2(g) + 1.5H_2O(l)$ |
| P | $P_4O_{10}(s)$ | $PH_3(g) + 2O_2(g) \rightarrow 0.25P_4O_{10}(s) + 1.5H_2O(l)$ |
| S | $SO_2(g)$ | $H_2S(g) + 1.5O_2(g) \rightarrow SO_2(g) + H_2O(l)$ |
| Cl | HCl(g) if H is present | $CH_2Cl_2(l) + O_2(g) \rightarrow CO_2(g) + 2HCl(g)$ |
| | or $Cl_2(g)$ if H is absent | $CCl_4(l) + O_2(g) \rightarrow CO_2(g) + 2Cl_2(g)$ |
| metals | metal oxides | $CuS(s) + 1.5O_2(g) \rightarrow CuO(s) + SO_2(g)$ |
| O | no additional products | $CO(g) + 0.5O_2(g) \rightarrow CO_2(g)$ |

Standard enthalpies of combustion for selected substances are given in table 3. Certain substances, such as water, carbon dioxide, halogens, noble gases and unreactive metals, do not react with molecular oxygen, so their $\Delta H_c^{\ominus}$ values are assumed to be 0 kJ mol$^{-1}$. Molecular nitrogen is a special case, as it reacts with oxygen only at very high temperatures and produces oxides with positive enthalpies of formation (table 1). Therefore, the $\Delta H_c^{\ominus}$ value for $N_2(g)$ is also assumed to be 0 kJ mol$^{-1}$.

**Table 3.** Standard enthalpies of combustion of selected substances at 298 K

| Substance | $\Delta H_c^{\ominus}$ / kJ mol$^{-1}$ | Substance | $\Delta H_c^{\ominus}$ / kJ mol$^{-1}$ |
|---|---|---|---|
| C(s, graphite) | −393.5 | $H_2(g)$ | −285.8 |
| C(s, diamond) | −395.4 | HCN(l) | −671.5 |
| $CH_4(g)$ | −891.1 | $H_2S(g)$ | −562.0 |
| $C_2H_2(g)$ | −1301.0 | $N_2(g)$ | 0 |
| $C_2H_4(g)$ | −1411.2 | $NH_3(g)$ | −382.8 |
| $C_2H_6(g)$ | −1560.7 | $N_2H_4(l)$ | −622.2 |
| $C_6H_6(l)$ | −3267.6 | $PH_3(g)$ | −1186.6 |
| $CH_3COOH(l)$ | −874.3 | S(s, rhombic) | −296.8 |
| CO(g) | −283.0 | S(s, monoclinic) | −297.1 |

Note that $\Delta H_c^{\ominus}(H_2(g)) = \Delta H_f^{\ominus}(H_2O(l))$, as the combustion of hydrogen and the formation of water under standard conditions are represented by the same thermochemical equation:

$$H_2(g) + 0.5O_2(g) \rightarrow H_2O(l) \quad \Delta H^{\ominus} = -285.8 \text{ kJ}$$

**Question**

4   State the equations that represent $\Delta H_c^{\ominus}$ for the following substances: a) S(s, rhombic); b) $C_2H_4(g)$; c) $CH_3COOH(l)$; d) HCN(l).

Like standard enthalpies of formation, standard enthalpies of combustion can be used for calculating the enthalpy changes for chemical reactions where both the reactants and products are flammable:

$$\Delta H^{\circ}(\text{reaction}) = \Sigma \Delta H_c^{\circ}(\text{reactants}) - \Sigma \Delta H_c^{\circ}(\text{products})$$

Note that the subtraction order of $\Delta H_c^{\circ}$ values (reactants minus products) is opposite to that for $\Delta H_f^{\circ}$ values (products minus reactants) – see figure 5.

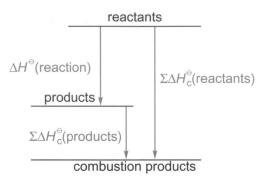

**Figure 5.** Calculating an enthalpy change from standard enthalpies of combustion

---

**Worked example: Calculating the enthalpy change from standard enthalpies of combustion**

**2.** Determine the $\Delta H^{\circ}$ value for the decomposition of hydrazine, $N_2H_4$.

$$N_2H_4(l) \rightarrow N_2(g) + 2H_2(g)$$

*Solution*

According to table 3, standard enthalpies of combustion for $N_2H_4(g)$, $N_2(g)$ and $H_2(g)$ are –622.2, 0 and –285.8 kJ mol$^{-1}$, respectively. Therefore:

$\Delta H^{\circ}(\text{reaction}) = -622.2 - (0 + 2 \times (-285.8)) = -50.6$ kJ.

---

**Question**

5   Using the data from table 3, calculate the standard enthalpy change for the following reactions:
    a) $C_2H_2(g) + 2H_2(g) \rightarrow C_2H_6(g)$; b) $2NH_3(g) \rightarrow N_2(g) + 3H_2(g)$.

6   Explain how your answer to part **(b)** of question 5 can be used to determine the standard enthalpy of formation of ammonia.

## Calorimetry

Enthalpies of combustion can be determined experimentally using a *bomb calorimeter* (figure 6). A sample of the substance to be studied is mixed with excess oxygen and ignited in a reaction chamber ("bomb") surrounded by water. The heat of the reaction raises the temperature of the water, which is measured with a very high precision (up to 0.0001 K) by an electronic thermometer. If we know the mass of the water ($m_w$), its specific heat capacity ($C_w = 4.18$ kJ kg$^{-1}$ K$^{-1}$) and the temperature change ($\Delta T$), then the amount of energy ($Q$) released by the combustion can be calculated as follows:

$$Q = m_w C_w \Delta T \tag{4}$$

In turn, the $\Delta H_c^{\circ}$ value of the substance can be found from $Q$, and the sample's mass ($m_s$) and its molar mass ($M_s$):

$$\Delta H_c^{\circ} = -\frac{M_s}{m_s} \times Q \tag{5}$$

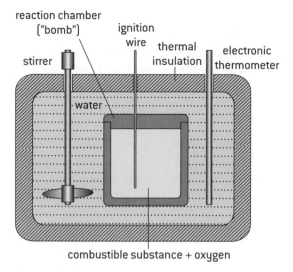

**Figure 6.** The bomb calorimeter

## Question

7  A sample of ethanol, $C_2H_5OH(l)$, with a mass of 0.288 g was burnt in a bomb calorimeter containing 4.00 kg of water. When the reaction was complete, the water temperature had increased from 298.000 to 298.511 K. Calculate the $\Delta H_c^\circ$ value for ethanol if the specific heat capacity of water is 4.18 kJ kg$^{-1}$ K$^{-1}$.

Thermochemical processes in aqueous solutions can be studied in the classroom using a *coffee-cup calorimeter*, which consists of a styrofoam (polystyrene foam) cup with a cork lid, a thermometer and a stirrer (figure 7). The cup material has a very low specific heat capacity and thermally insulates the reaction mixture from the surroundings. At the same time, styrofoam is resistant to most inorganic reagents, so aqueous solutions can be poured directly into the cup.

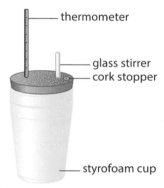

**Figure 7.** A coffee-cup calorimeter

## Practical skills

In a typical experiment with a coffee-cup calorimeter, each of the two reacting species is dissolved in a certain mass or volume of water, and both solutions are allowed to reach ambient temperature. The first solution is placed inside the calorimeter, and its temperature ($T_1$) is recorded. Then the second solution is added quickly through a small hole in the lid, the hole is plugged with a piece of cork, and the temperature of the reaction mixture is monitored until it reaches its maximum (for an exothermic reaction) or minimum (for an endothermic reaction). The maximum or minimum temperature ($T_2$) is also recorded and used for calculating the temperature change of the reaction mixture:

$$\Delta T = T_2 - T_1$$

If both reactant solutions are dilute, the specific heat capacity, mass and density of the reaction mixture will be approximately the same as those for pure water. In this case, the enthalpy change of the reaction can be calculated using equations (4) and (5).

A coffee-cup calorimeter can be used for determining the strength of acids and bases from the thermal effects of their neutralization reactions. Strong acids and bases are fully ionized in aqueous solutions, so their neutralization involves only $H^+(aq)$ and $OH^-(aq)$ ions:

$$H^+(aq) + OH^-(aq) \rightarrow H_2O(l)$$

**Internal link**

The strength of an acid or base is related to its degree of ionization in water, as discussed in **4.3 Solutions of electrolytes**.

Therefore, the neutralization of any strong acid with any strong base will have the same standard enthalpy change ($\Delta H_n^\circ = -57.6$ kJ mol$^{-1}$). In contrast, weak acids and bases undergo only partial ionization in aqueous solutions, as the formation of $H^+(aq)$ and $OH^-(aq)$ ions from their molecules is an endothermic process ($\Delta H^\circ > 0$). Thus, weak acids and bases will have smaller (less negative) enthalpies of neutralization, as some thermal energy will be used for breaking their molecules into ions.

Note that the experimental $\Delta H_n^\circ$ values obtained for any acid–base reaction in a coffee-cup calorimeter will be smaller than the theoretical value of $-57.6$ kJ mol$^{-1}$ because some of the heat of reaction will be lost to the surroundings or absorbed by the styrofoam cup, thermometer and stirrer. Such heat losses, known as *systematic errors*, cannot be totally eliminated but can be quantified, and, in some cases, reduced by altering the experimental conditions. For example, heat losses to the surroundings can be reduced by using two or three nested styrofoam cups, which will provide better thermal insulation. Other possible improvements might include the use of a larger volume of the solution, smaller thermometer and smaller stirrer. Finally, the heat capacity of the cup, thermometer and stirrer can be determined and added to the heat capacity of the reaction mixture in the calculation.

When the same coffee-cup calorimeter is used in a series of similar experiments, the systematic errors remain constant and reduce all measured enthalpies of neutralization by approximately the same amount. Therefore, strong and weak acids and bases can still be identified by the difference in their $\Delta H_n^\circ$ values.

### Worked example: Calculating the enthalpy change from temperature changes

**3.** In a coffee-cup calorimeter, 0.050 dm$^3$ of 1.0 M hydrochloric acid was mixed with 0.050 dm$^3$ of 1.0 M potassium hydroxide. Within seconds of mixing, the reaction mixture warmed up from 25.0 to 31.7°C and then began to cool down slowly. Calculate the standard enthalpy change for the reaction.

*Solution*

Since the initial solutions are dilute, the reaction mixture consists mostly of water, so it has approximately the same density (1.0 kg dm$^{-3}$) and specific heat capacity (4.18 kJ kg$^{-1}$ K$^{-1}$) as pure water. Each initial solution has a mass of approximately 0.050 dm$^3$ × 1.0 kg dm$^{-3}$ = 0.050 kg, so the total mass of the reaction mixture is 2 × 0.050 = 0.10 kg. Finally, the value for change in temperature $\Delta T$ in degrees Celsius is identical to that in kelvins, so:

$$\Delta T = 31.7 - 25.0 = 6.7°C = 6.7 \text{ K}$$

$$Q = m_w C_w \Delta T = 0.10 \text{ kg} \times 4.18 \text{ kJ kg}^{-1} \text{K}^{-1} \times 6.7 \text{ K} \approx 2.8 \text{ kJ}$$

Both hydrochloric acid and potassium hydroxide are strong electrolytes, so their neutralization can be represented as follows:

$$HCl(aq) + KOH(aq) \rightarrow KCl(aq) + H_2O(l)$$
$$H^+(aq) + OH^-(aq) \rightarrow H_2O(l)$$

The amount of each reactant is 0.050 dm³ × 1.00 mol dm⁻³ = 0.050 mol, so both hydrochloric acid and potassium hydroxide will be consumed completely, producing 0.050 mol of water. Therefore:

$$\Delta H_n^\circ = -\frac{2.8 \text{ kJ}}{0.050 \text{ mol}} = -56 \text{ kJ mol}^{-1}.$$

As expected, the experimental enthalpy of neutralization is slightly smaller (less negative) than the theoretical value of $-57.6$ kJ mol⁻¹. This difference is due to heat loss to the surroundings and other systematic errors introduced by the coffee-cup calorimeter.

## Question

8   When the experiment from worked example 3 was repeated using 0.050 dm³ of 1.0 M ethanoic acid ($CH_3COOH$) and 0.050 dm³ of 1.0 M potassium hydroxide, the temperature of the reaction mixture rose from 25.0 to 31.5°C. Calculate the standard enthalpy change for this neutralization reaction and suggest whether $CH_3COOH$ is a strong or weak acid.

## Bond enthalpy

Experimental determination of certain enthalpy changes can be problematic if the reactants are not readily available or the reaction proceeds too slowly under standard conditions. In such cases, the approximate $\Delta H^\circ$ value of the process can be found using the enthalpies of individual chemical bonds in all participating species (table 4).

**Table 4.** Average bond enthalpies (BE) at 298 K

| Bond | BE / kJ mol⁻¹ | Bond | BE / kJ mol⁻¹ |
|------|------|------|------|
| H–H | 436 | O–H | 463 |
| O–O | 144 | N–H | 391 |
| O=O | 498 | C–H | 414 |
| N–N | 158 | N–O | 214 |
| N=N | 470 | C–O | 358 |
| N≡N | 945 | C=O | 804 |
| C–C | 346 | C–N | 286 |
| C=C | 614 | C=N | 615 |
| C≡C | 839 | C≡N | 890 |

Each covalent bond is characterized by a certain *bond dissociation enthalpy* or simply *bond enthalpy* (*BE*), which is the energy required to break one mole of such bonds in a gaseous state under standard conditions. Bond breaking is an endothermic process, so all bond enthalpies are positive. Stronger bonds require more energy to break and thus have higher BE values. As you can see from table 4, double bonds are stronger than single bonds, and triple bonds are stronger than double bonds.

Bond enthalpies for some covalent bonds, such as H–H or O=O, are known with high accuracy, because they refer to specific compounds ($H_2(g)$ and $O_2(g)$, respectively). However, many other BEs listed in table 4 are average values that have been derived from experimental data for a wide range of covalent molecules. Therefore, the enthalpy changes of chemical reactions calculated from BEs are not very reliable and should be treated only as approximations.

> **Key term**
>
> **Bond dissociation enthalpy** (bond enthalpy, BE) is the energy required to break one mole of the bonds in a gaseous state under standard conditions.

Since energy is consumed when bonds are broken, and released when bonds are formed, the enthalpy of a chemical reaction can be calculated as follows:

$$\Delta H^{\circ}(\text{reaction}) = \Sigma BE(\text{bonds broken}) - \Sigma BE(\text{bonds formed})$$

However, instead of trying to identify specific bonds that are affected by a chemical change, it is convenient to assume that *all* bonds in the reactants are broken and *all* bonds in the products are formed during the reaction. According to Hess's Law, this assumption will not affect the overall enthalpy change, as the initial and final states of the system will remain unchanged:

$$\Delta H^{\circ}(\text{reaction}) = \Sigma BE(\text{reactants}) - \Sigma BE(\text{products})$$

**Worked example: Calculating the enthalpy change from bond enthalpies** WE

**4.** Calculate the standard enthalpy change for the following process using bond enthalpies:

$$N_2(g) + 3H_2(g) \rightarrow 2NH_3(g)$$

*Solution*

The reactants contain one N≡N bond (BE = 945 kJ mol$^{-1}$) in one molecule of nitrogen and three H–H bonds (BE = 436 kJ mol$^{-1}$) in three molecules of hydrogen:

$$\Sigma BE(\text{reactants}) = 945 + 3 \times 436 = 2253 \text{ kJ}$$

The products contain six N–H bonds (BE = 391 kJ mol$^{-1}$) in two molecules of ammonia:

$$\Sigma BE(\text{products}) = 6 \times 391 = 2346 \text{ kJ}$$

Therefore, the estimated enthalpy change of the reaction is 2253 – 2346 = –93 kJ. The actual $\Delta H^{\circ}$ value for this reaction is –91.8 kJ, which is very close to our estimate

**Question** Q

9  a)  Estimate the standard enthalpy of combustion of methane, $CH_4(g)$, using the bond enthalpies from table 4.

   b)  Suggest why this estimate differs significantly from the $\Delta H^{\circ}_c$ value given in table 3.

## Entropy

As mentioned in *3.2 Classification of chemical reactions*, many exothermic reactions ($\Delta H < 0$) proceed spontaneously because the formation of new chemical bonds in products releases enough energy to break chemical bonds in the reactants. Similarly, endothermic reactions ($\Delta H > 0$) are usually non-spontaneous and can proceed only when energy is supplied from an external source. However, these rules are not universal, so some exothermic reactions are non-spontaneous while some endothermic reactions are spontaneous. For example:

- solid ammonium nitrate dissolves spontaneously in water in a highly endothermic process, and the temperature of the reaction mixture may drop below zero:

$$NH_4NO_3(s) \rightarrow NH_4^+(aq) + NO_3^-(aq) \qquad \Delta H^{\circ} = +17.5 \text{ kJ}$$

- silver metal does not react with molecular oxygen, although the standard enthalpy of formation of silver(I) oxide is negative:

$$2Ag(s) + 0.5O_2(g) \xrightarrow{\times} Ag_2O(s) \qquad \Delta H^{\circ} = -31.1 \text{ kJ}$$

Therefore, as well as the thermal effect, there must be another factor that affects the spontaneity of chemical and physical processes. This factor is usually expressed in terms of *entropy* (S), which is a measure of the disorder in a system. Similar to the enthalpy change ($\Delta H$), every chemical reaction is characterized by the entropy change ($\Delta S$), which can be calculated from entropies of reactants and products. A positive entropy change ($\Delta S > 0$) drives the reaction forward and makes it more spontaneous while a decrease in entropy ($\Delta S < 0$) prevents the reaction from happening or makes it less spontaneous.

> **Key term**
>
> **Entropy (S)** is a measure of the disorder in a system.
>
> A positive entropy change ($\Delta S > 0$) makes the reaction more likely to happen. A negative entropy change ($\Delta S < 0$) makes the reaction less likely to happen.

**Question**

10 Under the same conditions, gases tend to have higher entropies than liquids, and liquids usually have higher entropies than solids. Similarly, aqueous ions have higher entropies than the same ions in a solid lattice. Suggest the sign of the entropy change for each of the following processes:

a) $H_2O(g) \rightarrow H_2O(s)$

b) $2H_2O_2(l) \rightarrow 2H_2O(l) + O_2(g)$

c) $NaCl(s) \rightarrow Na^+(aq) + Cl^-(aq)$

d) $NaCl(aq) + AgNO_3(aq) \rightarrow AgCl(s) + NaNO_3(aq)$

e) $NH_4NO_3(s) \rightarrow N_2O(g) + 2H_2O(l)$

f) $N_2(g) + 3H_2(g) \rightarrow 2NH_3(g)$

Enthalpy change and entropy change act independently and thus can have either the same or opposite effects on the reaction spontaneity. So if:

> **DP link**
>
> 15.2 Entropy and spontaneity

- $\Delta H < 0$ and $\Delta S > 0$, the reaction proceeds spontaneously under any conditions, because both factors favour the reaction

- $\Delta H > 0$ and $\Delta S < 0$, the reaction is non-spontaneous because neither factor favours the reaction.

When the changes in enthalpy and entropy have the same sign (either $\Delta H < 0$ and $\Delta S < 0$ or $\Delta H > 0$ and $\Delta S > 0$), the two factors compete with each other, and the reaction spontaneity depends on the absolute values of $\Delta H$ and $\Delta S$. At moderate temperatures (298 K or below), the enthalpy change usually wins the contest, which is the reason why most exothermic reactions are spontaneous and most endothermic reactions are non-spontaneous.

If, however, the entropy change is very large and the enthalpy change is very small, an endothermic reaction can become spontaneous, as happens when ammonium nitrate is mixed with water and its highly ordered crystals turn into the very disordered ions in solution.

Further discussion of entropy and spontaneity is beyond the scope of this book, as these concepts are studied only at the additional higher level of the IB Chemistry Diploma Programme. For now, it is sufficient to understand that the spontaneity of a process depends not only on its thermal effect but also on the entropy change associated with this process.

**DP link**

**6 Chemical kinetics**

**Figure 8.** Fireworks use fast chemical reactions whereas corrosion of metal constructions takes centuries

# 5.2 Chemical kinetics

Chemical reactions proceed at different speeds. Fast reactions, such as explosions, occur within milliseconds, while slow reactions, such as the formation of fossil fuels, take millions of years. However, the terms "fast" and "slow" are very vague, so chemists need to describe the speed of chemical changes more precisely. Therefore, it is convenient to define the *average reaction rate* ($v_{avr}$) as the change in concentration ($\Delta c$) of a reactant or product per unit time ($\Delta t$):

$$v_{avr} = \frac{|\Delta c|}{\Delta t}$$

Typically, reaction rates are measured in mol dm$^{-3}$ s$^{-1}$ although other units, such as mmol dm$^{-3}$ min$^{-1}$ or mol m$^{-3}$ h$^{-1}$, can also be used.

---

**Maths skills**

A reaction rate cannot be negative, so it must be calculated using the absolute value (modulus) of the concentration change. For example, if the concentration of species X in a reaction mixture decreases from 0.50 to 0.20 mol dm$^{-3}$ over 25 seconds, the average reaction rate with respect to that species is calculated as follows:

$$v_{avr}(X) = \frac{|0.20 - 0.50|\ \text{mol dm}^{-3}}{25\ \text{s}} = \frac{0.30\ \text{mol dm}^{-3}}{25\ \text{s}} = 0.012\ \text{mol dm}^{-3}\ \text{s}^{-1}$$

---

When two or more species participating in a chemical reaction have different stoichiometric coefficients, the reaction rates with respect to these species will also be different.

---

**Worked example: Calculating average reaction rates** **WE**

**5.** Consider the following reaction:

$$2N_2O(g) \rightarrow 2N_2(g) + O_2(g)$$

Under certain conditions, the concentration of nitrogen(I) oxide in the reaction mixture decreases by 0.20 mol dm$^{-3}$ over 10 seconds, producing 0.20 mol dm$^{-3}$ of nitrogen and 0.10 mol dm$^{-3}$ of oxygen over the same period of time. Calculate $v_{avr}$ for each of the three species involved in the reaction.

*Solution*

$$v_{avr}(N_2O) = \frac{0.20\ \text{mol dm}^{-3}}{10\ \text{s}} = 0.020\ \text{mol dm}^{-3}\ \text{s}^{-1}$$

$$v_{avr}(N_2) = \frac{0.20\ \text{mol dm}^{-3}}{10\ \text{s}} = 0.020\ \text{mol dm}^{-3}\ \text{s}^{-1}$$

$$v_{avr}(O_2) = \frac{0.10\ \text{mol dm}^{-3}}{10\ \text{s}} = 0.010\ \text{mol dm}^{-3}\ \text{s}^{-1}$$

---

As you can see, in worked example 5, the average reaction rate with respect to oxygen gas is half the rate with respect to nitrogen(I) oxide and nitrogen gas. To avoid such ambiguities, the rates for individual species can be divided by their stoichiometric coefficients. The resulting overall reaction rate is independent of the species used in calculations:

$$v_{avr} = \frac{v_{avr}(N_2O)}{2} = \frac{v_{avr}(N_2)}{2} = \frac{v_{avr}(O_2)}{1} = 0.010 \text{ mol dm}^{-3} \text{ s}^{-1}$$

## Question

11 Under certain conditions, ammonia can be oxidized to nitrogen(II) oxide:

$$4NH_3(g) + 5O_2(g) \rightarrow 4NO(g) + 6H_2O(g)$$

Over a period of 5 s, the concentration of nitrogen(II) oxide in the reaction mixture increased from 0 to $6.0 \times 10^{-3}$ mol dm$^{-3}$. Calculate:

a) the average reaction rates with respect to NO(g) and $O_2$(g);

b) the average rate of the reaction as a whole.

## Measuring reaction rate

Direct measurement of concentrations can be problematic, so $\Delta c$ values are often calculated from other experimental data, such as changes in pressure, volume or mass of the reaction mixture or a particular substance. For instance, the reaction of magnesium metal with hydrochloric acid can be investigated by measuring the volume of hydrogen gas released from the solution:

$$Mg(s) + 2HCl(aq) \rightarrow MgCl_2(aq) + H_2(g)$$

In a typical experiment, a carefully measured volume of hydrochloric acid with known concentration is placed into a conical flask, and a sample of magnesium is added. The flask is immediately closed with a rubber bung and connected to a gas syringe (figure 9). The gas volume is recorded at regular time intervals until the reaction is complete. The change in concentration of hydrochloric acid and the reaction rate are calculated using gas laws and reaction stoichiometry, as explained in worked example 6.

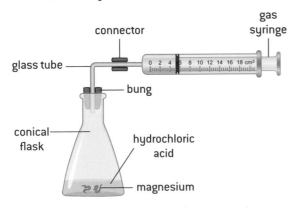

Figure 9. Measuring the reaction rate using a gas syringe

## Worked example: Calculating average reaction rates using gas laws

**WE**

**6.** The experiment shown in figure 9 was carried out using a piece of magnesium ribbon and $0.100$ dm$^3$ of $0.250$ M hydrochloric acid. The volume of gas released by the reaction was recorded every 10 seconds until a constant reading was obtained (table 5). Calculate the average reaction rates for the following time intervals: a) between 0 and 10 s; b) between 0 s and the moment when the reaction was complete. Assume that the reaction was carried out at SATP (25°C and 100 kPa), and the solution volume did not change during the reaction.

**Table 5.** Volume of hydrogen gas released by the reaction of hydrochloric acid with magnesium metal at SATP

| $t/s$ | 0 | 10 | 20 | 30 | 40 | 50 | 60 | 70 | 80 | 90 | 100 | 110 |
|---|---|---|---|---|---|---|---|---|---|---|---|---|
| $V(H_2)/cm^3$ | 0 | 111 | 164 | 192 | 210 | 223 | 234 | 241 | 246 | 248 | 248 | 248 |

*Solution*

a) Over the first 10 s of the reaction, 111 cm$^3$ (0.111 dm$^3$) of hydrogen was released at SATP. Under these conditions,

$V_m = 24.8$ dm$^3$ mol$^{-1}$,

so $n(H_2) = 0.111$ dm$^3$/24.8 dm$^3$ mol$^{-1} \approx 0.00448$ mol.

According to the equation, each mole of hydrogen requires two moles of hydrogen chloride, so:

$\Delta n(HCl) = 2 \times 0.00448$ mol $\approx 0.00896$ mol;

$\Delta c(HCl) = \dfrac{0.00896 \text{ mol}}{0.100 \text{ dm}^3} = 0.0896$ mol dm$^{-3}$;

$v_{avr} = \dfrac{0.0896 \text{ mol dm}^{-3}}{2 \times 10 \text{ s}} = 0.00448$ mol dm$^{-3}$ s$^{-1}$.

Note that the stoichiometric coefficient of HCl(aq) appears in the denominator of the last expression.

The same result could be obtained directly from the amount of hydrogen. Indeed, 0.00448 mol of $H_2$(g) was released from 0.100 dm$^3$ of the reaction mixture over 10 s, so:

$v_{avr} = \dfrac{0.00448 \text{ mol}}{0.100 \text{ dm}^3 \times 10 \text{ s}} = 0.00448$ mol dm$^{-3}$ s$^{-1}$.

b) According to table 5, the gas volume between 90 and 110 s remained constant, so the reaction was complete after approximately 90 s.

Over the same period, 248 cm$^3$ (0.248 dm$^3$) of hydrogen was released, so:

$n(H_2) = \dfrac{0.248 \text{ dm}^3}{24.8 \text{ dm}^3 \text{ mol}^{-1}} = 0.0100$ mol;

$\Delta n(HCl) = 2 \times 0.0100$ mol $= 0.0200$ mol;

$\Delta c(HCl) = \dfrac{0.0200 \text{ mol}}{0.100 \text{ dm}^3} = 0.200$ mol dm$^{-3}$;

$v_{avr} = \dfrac{0.200 \text{ mol dm}^{-3}}{2 \times 90 \text{ s}} = 1.11 \times 10^{-3}$ mol dm$^{-3}$ s$^{-1}$.

## Question

**12** Chalk (calcium carbonate) reacts with hydrochloric acid to release carbon dioxide:

$$CaCO_3(s) + 2HCl(aq) \rightarrow CaCl_2(aq) + CO_2(g) + H_2O(l)$$

A conical flask was charged with 50.0 cm³ of 0.500 M hydrochloric acid and a piece of chalk, plugged with cotton wool (to prevent droplets of water from escaping but let the gas through), placed on a digital balance and tared. The balance readings were recorded every 30 s using a data logger (table 6).

**Table 6.** Balance readings for the flask with the reaction mixture from question 12

| t / s | 0 | 30 | 60 | 90 | 120 | 150 | 180 | 210 | 240 | 270 | 300 |
|---|---|---|---|---|---|---|---|---|---|---|---|
| Δm / g | 0 | −0.161 | −0.262 | −0.326 | −0.368 | −0.396 | −0.415 | −0.428 | −0.437 | −0.440 | −0.440 |

Calculate the average reaction rates for the following time intervals: **a)** between 0 and 30 s; **b)** between 0 s and the moment when the reaction was complete. Assume that the solution volume did not change during the reaction.

## Instantaneous reaction rate

If you look closely at table 5, you will notice that the rate of the reaction between magnesium metal and hydrochloric acid changes with time. Indeed, the volume of hydrogen released during the first 10 s of the reaction (111 cm³) was more than 50 times greater than that produced during the last 10 s (2 cm³). Therefore, the average reaction rate gives us only a general idea of how fast (or slow) the reaction proceeds over a period of time but tells us nothing about the reaction rate at any given moment. To obtain this information, we need to introduce the concept of *instantaneous reaction rate* ($v_{inst}$), which is defined as the concentration change (d$c$) over an infinitesimally small period of time (d$t$):

$$v_{inst} - \frac{|dc|}{dt}$$

If an instantaneous rate is measured with respect to a particular substance, it must be divided by the stoichiometric coefficient of that substance to give the overall reaction rate.

### Key term

**Instantaneous reaction rate** ($v_{inst}$) is the instantaneous rate of change of concentration at a given moment, or in other words the concentration change over an infinitesimally small period of time.

**Initial reaction rate** ($v_{init}$) is the instantaneous rate measured at $t = 0$.

### Maths skills

A chemical equation can involve several species (indicated by the uppercase letters in the chemical equation below) with various stoichiometric coefficients (indicated by the lowercase letters):

$$aA + bB + \ldots \rightarrow pP + qQ + \ldots$$

IUPAC defines the instantaneous reaction rate for this general equation as follows:

$$v_{inst} = -\frac{1}{a}\frac{dc_A}{dt} = -\frac{1}{b}\frac{dc_B}{dt} = \frac{1}{p}\frac{dc_P}{dt} = \frac{1}{q}\frac{dc_Q}{dt}$$

Just as for $v_{avr}$, instantaneous reaction rate cannot be negative, so it must be calculated using absolute values of concentration changes. Note that reactants A and B are consumed during the reaction, so the changes of their concentrations are negative. To ensure that $v_{inst}$ is positive, the signs of these changes in the rate expression are reversed.

In contrast, the concentrations of products P and Q increase during the reaction, so their changes are positive and thus can be used in the rate expression without modification.

The instantaneous reaction rate at any given time $(t_x)$ can be determined by plotting the concentration of a reactant or product against time and drawing a tangent line to the curve at $t = t_x$. The slope (gradient) of the tangent line will be numerically equal to $v_{inst}$ at this time. Similarly the initial rate can be determined by measuring the slope of the tangent line at $t = 0$ s.

---

**DP ready**    **Theory of knowledge**

**Abstraction in science**

Instantaneous and initial reaction rates are mathematical abstractions, as they refer to changes in concentration over infinitesimally small periods of time. Nevertheless, these abstract quantities are used by chemists for various practical purposes, from optimizing reaction yields to deducing reaction mechanisms.

---

## Worked example: Determining initial and instantaneous reaction rates

**7.** Determine the initial rate and the instantaneous rate at $t = 55$ s for the reaction of hydrochloric acid with magnesium metal in worked example 6.

*Solution*

To do so, we need to find concentrations of hydrochloric acid at all experimental points.

The initial concentration of the acid (0.250 mol dm$^{-3}$) is given in the problem. At $t = 10$ s, $\Delta c(HCl)$ = 0.0896 mol dm$^{-3}$ (see worked example 6), so $c(HCl) = 0.250 - 0.0896 \approx 0.160$ mol dm$^{-3}$. All other concentrations up to $t = 90$ s can be calculated in the same way (table 7).

**Table 7.** Concentration of hydrochloric acid from worked example 6

| t / s | 0 | 10 | 20 | 30 | 40 | 50 | 60 | 70 | 80 | 90 |
|---|---|---|---|---|---|---|---|---|---|---|
| c(HCl) / mol dm$^{-3}$ | 0.250 | 0.160 | 0.118 | 0.095 | 0.081 | 0.070 | 0.061 | 0.056 | 0.052 | 0.050 |

Now we can plot these concentrations against time and draw tangent lines to the curve at $t = 0$ s and $t = 55$ s (figure 10). To determine the slope, we can select any two points on each tangent line and divide the difference in their $y$-coordinates $(\Delta c)$ by the difference in their $x$-coordinates $(\Delta t)$. For the tangent at $t = 0$ s, the most obvious point is $(x_0, y_0) = (0, 0.250)$ while the second point can be selected arbitrarily, for example, $(x_1, y_1) = (14, 0.100)$. This gives:

$$slope(0) = \frac{0.100 - 0.250}{14 - 0} \approx -0.0107$$

Since the stoichiometric coefficient before HCl(aq) is 2, the initial rate of the reaction will be half the absolute slope value:

$$v_{init} = \frac{|slope(0)|}{2} = \frac{|-0.0107|}{2} \approx 0.0054 \text{ mol dm}^{-3} \text{ s}^{-1}$$

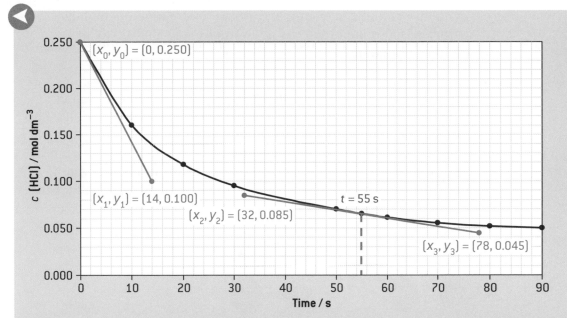

**Figure 10.** Determining initial and instantaneous reaction rates from experimental data

For the tangent at $t = 55$ s, two possible points are $(x_2, y_2) = (32, 0.085)$ and $(x_3, y_3) = (78, 0.045)$, so:

$$slope(55) = \frac{0.045 - 0.085}{78 - 32} \approx -8.7 \times 10^{-4}$$

$$v_{inst}(55) = \frac{|slope(55)|}{2} = \frac{|-8.7 \times 10^{-4}|}{2} \approx 4.4 \times 10^{-4} \text{ mol dm}^{-3} \text{ s}^{-1}$$

## Question

13 For all experimental points in question 12, calculate the concentrations of calcium chloride in the reaction mixture and plot these concentrations as a function of time on a graph paper. Using the tangent line method, determine the initial reaction rate and the instantaneous reaction rate at $t = 150$ s.

### Factors affecting rates of reaction

Reaction rates depend on many factors, such as temperature, concentrations of reactants, pressure (for gases) and the presence of a catalyst. Before discussing these factors in detail, we need to understand how chemical reactions occur at the molecular level. According to the *collision theory*, a chemical change can take place only when two (or sometimes more) reactant particles collide with each other, have correct mutual orientations and possess sufficient kinetic energy to initiate the reaction.

In gases and liquids, the molecules or ions move randomly and constantly collide with one another. Most of these collisions do not lead to any chemical changes, as the sum of kinetic energies of colliding particles is insufficient for breaking chemical bonds. As a result, the particles simply bounce off each other like billiard balls and fly in opposite directions. Such collisions are called *unsuccessful*, as they do not affect the chemical nature of the colliding particles.

Collisions redistribute kinetic energy between particles unequally, so some particles accelerate while others slow down. A collision between two fast-moving particles might be violent enough to break or rearrange chemical bonds and transform the reactants into products. Such collisions are called *successful*, as they lead to chemical changes.

**DP link**

6.1 Collision theory and rates of reaction

A.3 Catalysts

B.2 Proteins and enzymes

In addition to kinetic energy, the mutual orientation of colliding particles is also important. As an example, consider the following exchange reaction:

$$AB + CD \rightarrow AD + BC$$

For a collision to be successful, existing covalent bonds A–B and C–D must be broken, and new covalent bonds A–D and B–C must be formed. This is only possible if atom A comes into close proximity to atom D, and atom B comes into close proximity to atom C (figure 11, top). Collisions in other orientations will be unsuccessful, even if the kinetic energies of molecules AB and CD are high enough for the reaction to take place (figure 11, bottom).

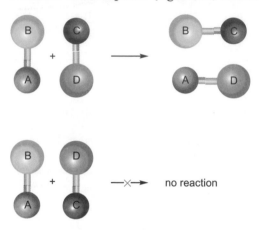

**Figure 11.** Favourable (top) and unfavourable (bottom) orientations of colliding molecules

Some reacting species, such as individual atoms and monoatomic ions, are symmetrical, so the results of their collisions do not depend on orientation. Chemical reactions between these species tend to proceed faster than reactions involving large and complex molecules, where the orientation of reactants is particularly important.

It follows from collision theory that the rate of a chemical reaction is proportional to the frequency of successful collisions in a given volume of the reaction mixture. Any change in reaction conditions that affects the number of collisions per second or the average kinetic energy of colliding particles will also affect the reaction rate. For example, an increase in **concentration** of a reactant will lead to more frequent collisions between the particles of this reactant and other species, so the reaction rate will increase. Conversely, a decrease in concentration of any reactant will slow down the reaction, as was demonstrated in worked examples 6 and 7.

**Pressure** affects the rates of reactions with gaseous reactants in the same way as concentration: an increase in pressure increases the reaction rate while a decrease in pressure has the opposite effect. Unlike gases, liquids and solids are almost incompressible, so pressure has no effect on reactions that do not involve gaseous reactants.

In heterogeneous mixtures, the collisions between reactant molecules are possible only at the surface where the different *phases* meet. Therefore, the rates of heterogeneous reactions depend on the **surface area** of reacting species. When a solid reactant is broken down into smaller pieces, its surface area increases, and so does the reaction rate. For instance, if in worked example 6 the magnesium ribbon were replaced with magnesium powder, the reaction would be complete within a few seconds. Similarly, when a reaction involves a liquid and

 **Internal link**

For a definition of heterogeneous mixtures, see **1.2 Chemical substances, formulae and equations**.

a gas, the reaction rate increases when the gas is bubbled through the liquid rather than allowed to pass over the liquid surface.

Heterogeneous reactions often proceed more slowly than homogeneous reactions, where successful collisions are possible in the whole volume of the reaction mixture. However, when the surface area of reactants is extremely large, heterogeneous reactions can also be very fast. Fine powders of active metals ignite spontaneously in air and react explosively with water and acids. Similarly, combustion of dispersed substances such as flour or coal dust can be very violent, which in the past led to a number of major explosions in grain mills and coal mines.

Concentration, pressure and surface area affect reaction rates by increasing or decreasing the frequency of collisions between reacting species. However, none of these factors affects the kinetic energy of colliding particles, so the proportion of successful collisions remains the same. This proportion can be altered, though, by changing the **temperature** of the reaction mixture.

When temperature increases, the average speed and kinetic energy of particles also increase. As particles move faster, they collide more often with one another. At the same time, a higher percentage of these collisions are successful, as more particles have sufficient kinetic energies for the reaction to occur. Therefore, the rate of almost any reaction increases with temperature. For most reactions at moderate temperatures (0–100°C), the rate approximately doubles when the temperature increases by 10°C. Conversely, low temperature slows down almost all chemical and biochemical reactions. For example, bacterial activity in food is slowed down at low temperatures, so refrigerated food remains fresh for much longer than that stored at room temperature.

> ### Key term
>
> A **phase** is an individual substance or mixture that has uniform chemical and physical properties. Although the term "phase" is often used as a synonym for "state of matter", immiscible liquids can form two or more separate phases with different chemical compositions. Similarly, each solid substance in a heterogeneous mixture is a separate phase.

## Question

14 Hydrogen peroxide in aqueous solutions decomposes as follows:

$$2H_2O_2(aq) \rightarrow 2H_2O(l) + O_2(g)$$

The graph in figure 12 shows the volume of oxygen gas released in this reaction as a function of time.

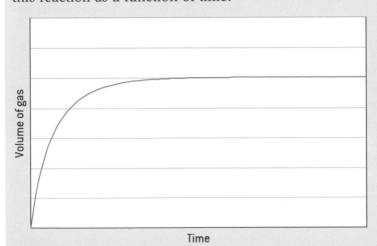

**Figure 12.** Volume of oxygen gas released by decomposition of aqueous hydrogen peroxide over time

On the same graph, sketch and label three curves for the following changes in experimental conditions: **a)** increased initial concentration of hydrogen peroxide; **b)** lowered temperature of the reaction mixture; **c)** increased atmospheric pressure.

## Key term

A **catalyst** is a substance that increases the rate of a chemical reaction but is not consumed in that reaction.

An **intermediate** is an unstable compound formed between a catalyst and a reactant. This intermediate goes on to form the reaction product.

**Activation energy** is the minimum energy required for the reaction to occur.

## Catalysis

In addition to raising the temperature, the proportion of successful collisions in a reaction mixture can be increased by the use of a *catalyst*. Typically, a catalyst forms an unstable compound (an *intermediate*) with a reactant. This intermediate then undergoes further chemical changes and eventually forms the reaction product, releasing the catalyst in an unchanged form. Therefore, the catalyst itself is both a reactant and product of the same reaction.

The effect of a catalyst on the reaction rate can be explained in terms of *activation energy* ($E_a$), which is the minimum energy required for the reaction to occur.

- When the sum of kinetic energies of colliding particles is less than the activation energy ($E_{kin} < E_a$), the collision between these particles will be unsuccessful, regardless of their orientation.

- When $E_{kin} \geq E_a$, the collision can be successful if the orientation is correct.

An increase in temperature does not affect $E_a$ but increases the average $E_{kin}$, so a greater proportion of collisions become successful. In contrast, a catalyst does not affect the average $E_{kin}$ but reduces $E_a$ by providing an alternative pathway for the reaction (figure 13). A lower $E_a$ means that more particles will have enough energy to react with one another, so the reaction rate will increase.

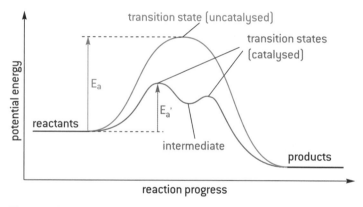

**Figure 13.** Typical energy profiles for uncatalysed and catalysed reactions

## Key term

**Homogeneous catalysis** takes place when the catalyst and reactants form a single phase, while **heterogeneous catalysis** occurs when the catalyst and reactants are present in different phases.

An important case of *homogeneous catalysis* is the decomposition of stratospheric ozone, outlined in the **Nature of science** box on the following page. *Heterogeneous catalysis* is used in the synthesis of ammonia (see *5.3 Chemical equilibrium*) and many other industrial processes. An efficient catalyst may accelerate the reaction by several orders of magnitude and thus greatly reduce the cost and production time of the target compound.

## Ozone depletion

Stratospheric ozone absorbs high-energy ultraviolet (UV) radiation from the Sun and thus protects all living organisms from its harmful effects. Chlorine-containing substances, such as chlorofluorocarbons (CFCs), reduce the concentration of ozone by catalysing its decomposition. Various CFCs were formerly used in air-conditioning systems, refrigerators and aerosol cans, so large quantities of these substances were released into the atmosphere. The C–Cl bonds in CFCs are easily broken by UV light, producing very reactive chlorine radicals, Cl ·. These radicals eventually reach the stratosphere and break down ozone, $O_3$, into diatomic oxygen, $O_2$, as follows:

$$O_3(g) + Cl \cdot (g) \rightarrow O_2(g) + ClO \cdot (g)$$
$$O_3(g) + ClO \cdot (g) \rightarrow 2O_2(g) + Cl \cdot (g)$$

Initially, a chlorine radical attacks a molecule of ozone, producing a molecule of diatomic oxygen and another radical, $ClO \cdot (g)$. This radical is unstable and quickly reacts with another ozone molecule, producing more diatomic oxygen and regenerating the chlorine radical. The activation energy of this two-step process is very low, so $Cl \cdot (g)$ acts as a homogeneous catalyst, increasing the rate of ozone decomposition without being consumed in the process. Another radical, $ClO \cdot (g)$, is the reaction intermediate (figure 13).

 **Internal link**

Ozone is introduced in **2.3 Chemical bonding**.

## Question

15 Sulfur(IV) oxide reacts with molecular oxygen as follows:

$$2SO_2(g) + O_2(g) \rightarrow 2SO_3(g)$$

In the presence of nitrogen(IV) oxide, the rate of this reaction increases up to 10,000 times owing to the following two-step process:

$$SO_2(g) + NO_2(g) \rightarrow SO_3(g) + NO(g)$$
$$NO(g) + 0.5O_2(g) \rightarrow NO_2(g)$$

State the roles of nitrogen oxides in this process and explain their effects on the reaction rate.

Biochemical reactions are assisted by protein-based catalysts (*enzymes*). Without enzymatic catalysis, many processes in living organisms would be impossible and life itself would not exist (at least not in its present form).

**Key term**

**Enzymes** are biological, protein-based catalysts.

## 5.3 Chemical equilibrium

Reversible reactions can proceed in both directions at the same time. A typical example of a reversible process is the reaction involving gaseous nitrogen, hydrogen and ammonia:

$$N_2(g) + 3H_2(g) \rightleftharpoons 2NH_3(g)$$

The equilibrium sign ($\rightleftharpoons$) in the above equation means that both processes—the synthesis of ammonia from nitrogen and

**Internal link**

The concept of reversible reactions is introduced in **3.2 Classification of chemical reactions**.

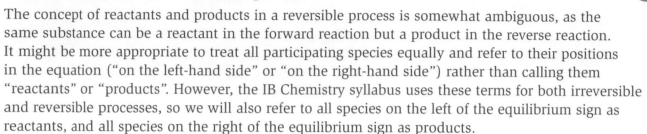
hydrogen (*forward reaction*) and the decomposition of ammonia into nitrogen and hydrogen (*backward*, or *reverse reaction*)—take place simultaneously, so reactant and product molecules are constantly interconverted.

**Reactants and products in reversible processes**

The concept of reactants and products in a reversible process is somewhat ambiguous, as the same substance can be a reactant in the forward reaction but a product in the reverse reaction. It might be more appropriate to treat all participating species equally and refer to their positions in the equation ("on the left-hand side" or "on the right-hand side") rather than calling them "reactants" or "products". However, the IB Chemistry syllabus uses these terms for both irreversible and reversible processes, so we will also refer to all species on the left of the equilibrium sign as reactants, and all species on the right of the equilibrium sign as products.

The predominant direction of a reversible reaction depends on the initial concentrations of all participating species. If, for example, the initial mixture contains only nitrogen and hydrogen, only the forward reaction takes place: some molecules of nitrogen and hydrogen combine together to form ammonia. Over time, the concentrations of nitrogen and hydrogen in the reaction mixture decrease while the concentration of ammonia increases (figure 14(a)).

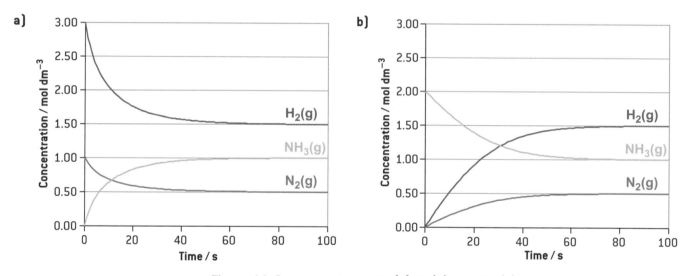

**Figure 14.** Concentrations of $N_2(g)$, $H_2(g)$ and $NH_3(g)$ in reaction mixtures approaching equilibrium at 475 K

**Key term**

**Chemical equilibrium** is a dynamic state where two or more ongoing processes perfectly balance one another, so each participating species is consumed and produced at the same rate. The apparent absence of chemical changes at equilibrium is only an illusion, as all components of the reaction mixture are constantly transformed into one another.

Once the molecules of ammonia appear in the reaction mixture, the reverse reaction becomes possible, so these ammonia molecules begin to decompose into nitrogen and hydrogen. This process accelerates as the concentration of ammonia increases. At the same time, the forward reaction slows down, as the mixture contains fewer and fewer molecules of nitrogen and hydrogen. Eventually, the rates of forward and reverse reactions become exactly the same, and the concentrations of all three substances (nitrogen, hydrogen and ammonia) remain constant even though the reactions continue. This state is known as *chemical equilibrium*.

Chemical equilibrium can be achieved from any initial state of the system. In figure 14(a), the initial mixture contained 1.00 mol dm$^{-3}$ of nitrogen, 3.00 mol dm$^{-3}$ of hydrogen and no ammonia, while the equilibrium concentrations were 0.50, 1.50 and 1.00 mol dm$^{-3}$, respectively. However, exactly the same equilibrium concentrations of all three species will be produced if we start the reaction with 2.00 mol dm$^{-3}$ of ammonia only, with no nitrogen or hydrogen in the initial mixture (figure 14(b)). In this case, some ammonia will decompose into nitrogen and hydrogen:

$$2NH_3(g) \rightleftharpoons N_2(g) + 3H_2(g)$$

As the process continues, the concentration of ammonia will decrease while the concentrations of nitrogen and hydrogen will increase. Eventually, the rates of forward and reverse reactions will become equal, and equilibrium will be reached.

## Equilibrium constant

The position of chemical equilibrium can be characterized by the *equilibrium constant* ($K_c$), which is the ratio of the equilibrium concentrations of reactants and products raised to the power of their stoichiometric coefficients. For example, the $K_c$ expression for the synthesis of ammonia from nitrogen and hydrogen will look as follows:

$$N_2(g) + 3H_2(g) \rightleftharpoons 2NH_3(g) \qquad K_c = \frac{[NH_3]^2}{[N_2][H_2]^3}$$

Note that the equilibrium concentrations are denoted by square brackets around the chemical formulae, and the states of substances in $K_c$ expressions are often omitted. For convenience, $K_c$ values are generally treated as having no units.

> **Key term**
>
> The **equilibrium constant** (**$K_c$**) is the ratio of the equilibrium concentrations of products over reactants, each raised to the power of their stoichiometric coefficient.

### Maths skills: Writing $K_c$ expressions

Consider the general case of a chemical equilibrium involving several species with various stoichiometric coefficients:

$$aA + bB + \dots \rightleftharpoons pP + qQ + \dots$$

In the $K_c$ expression, the products of the forward process (P, Q, …) appear in the numerator while the reactants of the forward process (A, B, …) appear in the denominator:

$$K_c = \frac{[P]^p[Q]^q \dots}{[A]^a[B]^b \dots}$$

The use of square brackets and the omission of state symbols saves space and focuses our attention on the most important information about the system at equilibrium. For the same reason, $K_c$ values are treated as unitless quantities, although all equilibrium concentrations must still be expressed using appropriate units (typically, mol dm$^{-3}$).

**Worked example: Calculating the equilibrium constant**

**8.** Calculate the $K_c$ value for the synthesis of ammonia at 475 K, using the data from figure 14.

*Solution*

$$K_c = \frac{1.00^2}{0.50 \times 1.50^3} \approx 0.59$$

**Question**

16 a) State the $K_c$ expression for the following equation:

$$2NH_3(g) \rightleftharpoons N_2(g) + 3H_2(g)$$

b) Using the data from figure 14, calculate the $K_c$ value for this process.

For any given process, the value of the equilibrium constant depends only on temperature and not on other reaction conditions, such as pressure or concentration. For example, if we increase the initial concentration of $N_2(g)$ from 1.0 to 2.5 mol dm⁻³, the state of equilibrium will be achieved at different concentrations of nitrogen, hydrogen and ammonia (figure 15). However, these new concentrations (1.85, 1.13 and 1.25 mol dm⁻³, respectively) will still satisfy the $K_c$ expression:

$$K_c = \frac{1.25^2}{1.85 \times 1.13^3} \approx 0.59$$

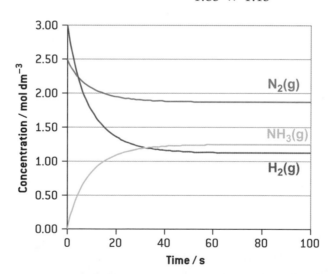

**Figure 15.** Concentrations of $N_2(g)$, $H_2(g)$ and $NH_3(g)$ in a reaction mixture approaching equilibrium at 475 K

Therefore, the $K_c$ value can be used for determining the composition of the reaction mixture at equilibrium, as shown in worked example 9.

$K_c$ and spontaneity

The value of $K_c$ provides important information about the direction
and extent of the spontaneous process under standard conditions.
If $K_c > 1$, the products are favoured over the reactants, so the
forward reaction will proceed spontaneously until most reactants
are converted into products. Conversely, if $K_c < 1$, the reactants
are favoured, so the reverse reaction will be spontaneous. Finally,
if $K_c \approx 1$, the equilibrium will be established at approximately equal
concentrations of reactants and products.

**Worked example: Calculating concentrations using the equilbrium constant**

**WE**

**9.** A mixture of sulfur(IV) oxide and oxygen was heated in a sealed vessel at 1000 K until the
following equilibrium was reached:

$$2SO_2(g) + O_2(g) \rightleftharpoons 2SO_3(g)$$

The equilibrium concentrations of sulfur(IV) oxide and sulfur(VI) oxide in the final mixture were
0.12 and 0.18 mol dm$^{-3}$, respectively. Calculate the equilibrium concentration of oxygen and initial
concentrations of both reactants if the $K_c$ value for this reaction at 1000 K is 3.0.

*Solution*

First of all, we need to write the $K_c$ expression:

$$K_c = \frac{[SO_3]^2}{[O_2][SO_2]^2}$$

Since we know three of the four values in this expression, we can find the equilibrium
concentration of oxygen:

$$3.0 = \frac{0.18^2}{[O_2] \times 0.12^2};$$

$$[O_2] = 0.75 \text{ mol dm}^{-3}.$$

To find the initial concentrations of $SO_2(g)$ and $O_2(g)$, we need to look at the reaction stoichiometry.
The formation of each mole of $SO_3(g)$ consumes 1 mol of $SO_2(g)$ and 0.5 mol of $O_2(g)$, so:

$$c(SO_2(g))_{init} = [SO_2] + [SO_3] = 0.12 + 0.18 = 0.30 \text{ mol dm}^{-3};$$

$$c(O_2(g))_{init} = [O_2] + 0.5[SO_3] = 0.75 + 0.09 = 0.84 \text{ mol dm}^{-3}.$$

**Question**

**Q**

17 Nitrogen dioxide was cooled down to 10°C in a sealed vessel until it reached equilibrium with
dinitrogen tetroxide:

$$2NO_2(g) \rightleftharpoons N_2O_4(g) \qquad K_c = 11.5$$

Calculate the equilibrium and initial concentrations of $NO_2(g)$ if the equilibrium concentration of
$N_2O_4(g)$ was 0.041 mol dm$^{-3}$.

**Reaction quotient**

If a system has not reached equilibrium, the ratio of actual
concentrations of reacting species differs from the $K_c$ value. This ratio,
known as the *reaction quotient* ($Q_c$), can be used for determining the
direction of ongoing chemical changes within the system.

- When $Q_c < K_c$, the reaction mixture contains more reactants and
  less products than needed at equilibrium, so the forward reaction
  will be favoured.

 **Key term**

The **reaction quotient ($Q_c$)**
is calculated just like the $K_c$
expression, but with non-
equilibrium concentrations, and
can be used for determining the
direction of a reversible reaction
before it has reached equilibrium.

- When $Q_c > K_c$, there are less reactants and more products than needed, so the reverse reaction will be the dominant process.
- When $Q_c = K_c$, the system is already at equilibrium, so the forward and reverse reactions will proceed at the same rate.

## Worked example: Using the reaction quotient

**WE**

**10.** Calculate the reaction quotient for a mixture of nitrogen, hydrogen and ammonia at 475 K where $K_c = 0.59$ and the concentration of each species is 0.5 mol dm⁻³. Predict the direction of the favoured process in this mixture.

$$N_2(g) + 3H_2(g) \rightleftharpoons 2NH_3(g) \qquad K_c = \frac{[NH_3]^2}{[N_2][H_2]^3} = 0.59$$

*Solution*

To calculate the reaction quotient, we need to use the $K_c$ expression with actual concentrations instead of equilibrium concentrations:

$$Q_c = \frac{0.5^2}{0.5 \times 0.5^3} = 4$$

Since $Q_c > K_c$, the system contains too much ammonia and too little nitrogen and hydrogen, so the reverse reaction is favoured.

## Question

**Q**

18 Consider the reaction in question 17, where the reaction mixture contains 0.025 mol dm⁻³ of nitrogen dioxide and 0.10 mol dm⁻³ of dinitrogen tetroxide. Determine the direction of the favoured process in this mixture at 10°C ($K_c = 11.5$).

**Figure 16.** Henry Le Châtelier (1850–1936)

## Le Châtelier's principle

Chemical equilibrium is a dynamic process, so it can be easily disturbed by any change in the reaction conditions, such as temperature, pressure or concentrations of reacting species. The effects of such changes can be predicted by using *Le Châtelier's principle*, which was formulated at the end of the 19th century by the French chemist Henry Le Châtelier.

### Key term

**Le Châtelier's principle**: If a dynamic equilibrium is disturbed by a change in the reaction conditions, the balance between the forward and reverse processes will shift to counteract the change and return the system to equilibrium.

To illustrate Le Châtelier's principle, let's examine an equilibrium mixture of nitrogen, hydrogen and ammonia at 475 K where $[N_2] = 0.5$ mol dm⁻³, $[H_2] = 1.5$ mol dm⁻³, $[NH_3] = 1.0$ mol dm⁻³ and $K_c = 0.59$ (figure 17). Initially, the concentrations of all three gases remain constant, as the mixture is already at equilibrium ($Q_c = K_c$), and the rates of the forward and reverse reactions are equal:

$$N_2(g) + 3H_2(g) \rightleftharpoons 2NH_3(g)$$

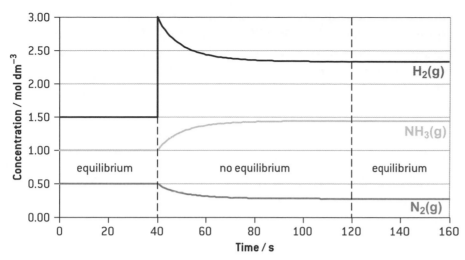

**Figure 17.** The effect of a concentration change on the equilibrium position

At $t$ = 40 s, we disturb the equilibrium by injecting more hydrogen into the reaction mixture, so the concentration of $H_2(g)$ increases from 1.5 to 3.0 mol dm$^{-3}$. This change in concentration leads to more frequent collisions between hydrogen and nitrogen molecules and thus increases the rate of the forward reaction. As a result, the concentrations of hydrogen and nitrogen in the reaction mixture decrease while the concentration of ammonia increases. In other words, the system counteracts the presence of excess reactant (hydrogen) by converting it into product (ammonia) and shifting the equilibrium position to the right. These changes do not affect the $K_c$ value, so the new equilibrium concentrations of nitrogen, hydrogen and ammonia (0.28, 2.33 and 1.45 mol dm$^{-3}$, respectively) still satisfy the $K_c$ expression:

$$K_c = \frac{1.45^2}{0.28 \times 2.33^3} \approx 0.59$$

The opposite effect would be observed if we increased the concentration of ammonia instead of hydrogen. Since ammonia is the product, its higher concentration would accelerate the reverse reaction and shift the equilibrium position to the left. In this case, the system would counteract the presence of excess product (ammonia) by converting it into reactants (nitrogen and hydrogen).

## Question

19  Outline how the following changes will affect the equilibrium position and the $K_c$ value of the reaction $N_2(g) + 3H_2(g) \rightleftharpoons 2NH_3(g)$:

a) increase in concentration of nitrogen; b) decrease in concentration of ammonia.

The effect of pressure on the equilibrium position depends on the stoichiometric ratio of gaseous reactants and products. In our example, there are four gaseous molecules (one $N_2$ and three $H_2$) on the left-hand side of the equation but only two gaseous molecules of $NH_3$ on the right-hand side. If the pressure increases, the system will counteract this change by converting the reactants into the products and thus reducing the number of gaseous molecules in the reaction mixture. Therefore, the equilibrium position will shift to the right. Conversely, a decrease in pressure will move the equilibrium position to the left, towards the greater number of gaseous molecules. In both cases, the $K_c$ value of the reaction will not change.

**Internal link**

The ideal gas law is described in **1.3 Stoichiometric relationships**.

## Maths skills: Pressure and volume

According to the ideal gas law, the pressure of a gas mixture is inversely proportional to its volume. When the volume of a system decreases, the pressure increases, and when the volume increases, the pressure decreases. Therefore, the effects of a change in volume on the equilibrium position are opposite to the effects of pressure. Neither volume nor pressure has any effect on the $K_c$ value.

If an equation contains more gaseous molecules on the right-hand side than on the left-hand side, an increase in pressure will shift the equilibrium position to the left while a decrease in pressure will shift the equilibrium position to the right. Finally, if the numbers of gaseous molecules on each side of the equation are equal, the pressure will have no effect on the equilibrium position. Once again, the $K_c$ value will remain unchanged in all cases.

## Question

20  State the effects of increasing pressure on the equilibrium positions and $K_c$ values for following reactions:

a) $N_2O_4(g) \rightleftharpoons 2NO_2(g)$

b) $SO_2(g) + NO_2(g) \rightleftharpoons SO_3(g) + NO(g)$

c) $4HCl(g) + O_2(g) \rightleftharpoons 2H_2O(g) + 2Cl_2(g)$

Temperature is the only factor that affects both the position of equilibrium and the equilibrium constant. The synthesis of ammonia from nitrogen and hydrogen is an exothermic process ($\Delta H^\circ < 0$), so heat is released by the forward reaction and consumed by the reverse reaction:

$$N_2(g) + 3H_2(g) \rightleftharpoons 2NH_3(g) \qquad \Delta H^\circ = -91.8 \text{ kJ}$$

At equilibrium, the forward and reverse reactions proceed at the same rates, so the total amount of heat in the system remains constant. If we increase the temperature, more heat will be introduced into the system. In accordance with Le Châtelier's principle, the system will counteract this change by consuming excess heat, so the reverse reaction will be favoured, and the equilibrium position will shift to the left. As a result, the concentration of ammonia in the reaction mixture will decrease while the concentrations of both nitrogen and hydrogen will increase.

Now let's look at the $K_c$ expression of the reaction:

$$K_c = \frac{[NH_3]^2}{[N_2][H_2]^3}$$

The numerator contains $[NH_3]$, which decreases with temperature. The denominator contains $[N_2]$ and $[H_2]$, both of which increase with temperature. Therefore, an increase in temperature lowers the $K_c$ value of this reaction. The same result will be observed for any other exothermic process. Conversely, a decrease in temperature shifts the equilibrium position of an exothermic process to the right and thus increases its $K_c$ value.

For an endothermic process, the reverse is true: an increase in temperature will shift the equilibrium position to the right and increase the $K_c$ value. A decrease in temperature will have opposite effects, shifting the equilibrium position to the left and decreasing the $K_c$ value.

Chemical or physical changes with $\Delta H^\circ = 0$ are very rare but still possible. For such processes, temperature has no effect on the equilibrium position or the $K_c$ value.

---

**DP ready**   **Theory of knowledge**

### Heat as a reactant or product

To predict the effect of a temperature change on the equilibrium position, it is convenient to treat heat as an imaginary substance ($Q$) participating in the reaction. For an exothermic process, this "substance" will act like a reaction product, and for an endothermic process, it will act like a "reactant":

$$\text{reactants} \rightleftharpoons \text{products} + Q \qquad \Delta H^\circ < 0 \text{ (exothermic)}$$

$$\text{reactants} + Q \rightleftharpoons \text{products} \qquad \Delta H^\circ > 0 \text{ (endothermic)}$$

An increase in temperature increases the "concentration" of heat in the system, so the equilibrium position of the first (exothermic) reaction will shift to the left while the equilibrium position of the second (endothermic) reaction will shift to the right. Similarly, a decrease in temperature decreases the "concentration" of heat and thus will have opposite effects on the above equilibria.

---

### Question

21 State the effects of decreasing temperature on the equilibrium positions and $K_c$ values for the following reactions:

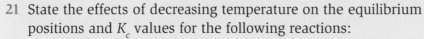

a) $N_2(g) + O_2(g) \rightleftharpoons 2NO(g)$      $\Delta H^\circ = +182.6 \text{ kJ}$

b) $2SO_2(g) + O_2(g) \rightleftharpoons 2SO_3(g)$      $\Delta H^\circ = -197.8 \text{ kJ}$

As was explained in *5.2 Chemical kinetics*, the rate of a chemical reaction increases in the presence of a catalyst. The catalyst provides an alternative pathway for the reaction and thus lowers the reaction's activation energy. In a reversible process, the forward and reverse reactions follow the same pathway in opposite directions, so the rates of both reactions increase to the same extent. Therefore, in the presence of a catalyst, the equilibrium state of the system is achieved faster, but the position of this equilibrium and the $K_c$ value do not change in any way.

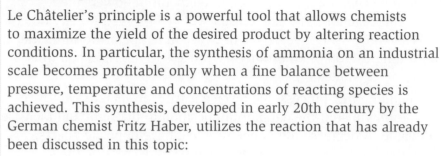

H₂(g)

N₂(g)

Compressor

Return pump

Reaction chamber

Condenser

NH₃(l)

**Figure 18.** Flow chart of the Haber process

## The Haber process

Le Châtelier's principle is a powerful tool that allows chemists to maximize the yield of the desired product by altering reaction conditions. In particular, the synthesis of ammonia on an industrial scale becomes profitable only when a fine balance between pressure, temperature and concentrations of reacting species is achieved. This synthesis, developed in early 20th century by the German chemist Fritz Haber, utilizes the reaction that has already been discussed in this topic:

$$N_2(g) + 3H_2(g) \rightleftharpoons 2NH_3(g) + Q$$

Since the number of gas molecules in the forward reaction decreases from four to two, the Haber process is carried out at high pressure (20 MPa, or 200 atm), which pushes the equilibrium position to the right and at the same time increases the reaction rate. Nitrogen and hydrogen are constantly injected into the reaction mixture while ammonia is condensed and removed after each cycle of the process (Figure 18). These measures allow for maximizing the concentrations of reactants and minimizing the concentration of the product, shifting the equilibrium even further to the right.

The forward reaction is exothermic ($\Delta H^\ominus < 0$) and thus is favoured by a low temperature. However, low temperature reduces the kinetic energy and frequency of collisions between reactant particles and thus decreases the reaction rate. As a compromise, the reaction is carried out at a moderate temperature (400–450°C) and in the presence of a catalyst (iron powder with various additives). While the catalyst itself does not affect the equilibrium position, it lowers the temperature required for the reaction and greatly increases the reaction rate, so more ammonia can be synthesized in the same apparatus per unit time.

In modern industrial plants, the efficiency of the Haber process is improved by recycling unreacted gases and utilizing the heat released by the forward reaction. Such improvements increase the overall reaction yield to 98% while reducing the cost and environmental impact of ammonia production.

Ammonia is widely used for making urea, ammonium nitrate and other fertilizers, without which current farming practices could not feed the growing world population. According to some estimates, well over half of the nitrogen in our bodies comes from the Haber process, making it arguably the most important invention of the 20th century. However, the process was also vital for the production of explosives and ammunition in both world wars. Haber himself also worked on and promoted the use of poison gas during the First World War, earning him the name "the father of chemical warfare". His story illustrates how advances in scientific knowledge can bring great benefit or do terrible harm.

**Figure 19.** Fritz Haber (1868–1934)

So far, we have considered only homogeneous equilibria involving gaseous reactants and products. Before we look at other reversible processes, let's summarize the effects of various factors on the equilibrium position and $K_c$ of reactions in the gas phase (table 8).

**Table 8.** The effects of reaction conditions on the equilibrium position and $K_c$ value

| Change in condition | Shift of equilibrium | $K_c$ |
|---|---|---|
| $c$(reactant) increases | to the right (towards products) | no change |
| $c$(product) decreases | | |
| $c$(reactant) decreases | to the left (towards reactants) | |
| $c$(product) increases | | |
| pressure increases | to the side with a smaller number of gas molecules | |
| volume decreases | | |
| pressure decreases | to the side with a greater number of gas molecules | |
| volume increases | | |
| temperature increases | $\Delta H° < 0$: to the left<br>$\Delta H° > 0$: to the right<br>$\Delta H° = 0$: no change | $\Delta H° < 0$: decreases<br>$\Delta H° > 0$: increases<br>$\Delta H° = 0$: no change |
| temperature decreases | $\Delta H° < 0$: to the right<br>$\Delta H° > 0$: to the left<br>$\Delta H° = 0$: no change | $\Delta H° < 0$: increases<br>$\Delta H° > 0$: decreases<br>$\Delta H° = 0$: no change |
| catalyst is added | no change | no change |

Not surprisingly, the same principles can be applied to all other reversible processes, from physical changes (such as evaporation or condensation) to chemical reactions in solutions (such as neutralization or precipitation). Let's consider the following equilibria:

$$H_2O(l) \rightleftharpoons H_2O(g) \qquad \Delta H° = +44.0 \text{ kJ} \quad K_c = [H_2O(g)]$$

$$NH_3(aq) + H^+(aq) \rightleftharpoons NH_4{}^+(aq) \qquad \Delta H° = -51.9 \text{ kJ} \quad K_c = \frac{[NH_4^+]}{[NH_3][H^+]}$$

$$Ca^{2+}(aq) + 2F^-(aq) \rightleftharpoons CaF_2(s) \qquad \Delta H° = -14.3 \text{ kJ} \quad K_c = \frac{1}{[Ca^{2+}][F^-]^2}$$

All these processes obey Le Châtelier's principle, so we can predict the effects of temperature, pressure and concentrations on their equilibrium positions using table 8. At the same time, each process can be characterized by an equilibrium constant, $K_c$, which depends only on temperature but not on other reaction conditions. Note that concentrations of solid and liquid species are assumed to be constant, so they do not appear in $K_c$ expressions. Also, if an equation involves the same substance in different states, such as $H_2O(l)$ and $H_2O(g)$ in the first example, these states must be shown in the $K_c$ expression.

## Chapter summary

In this chapter, you have learned about thermochemistry, chemical kinetics and equilibrium. Before moving further, make sure that you have a working knowledge of the following concepts and definitions:

☐ Enthalpy ($H$) is the total heat content of a system.

☐ In a closed system at constant pressure, $\Delta H = -Q$

☐ For an exothermic reaction $\Delta H < 0$, while for an endothermic reaction $\Delta H > 0$.

- ☐ The thermal effect of a chemical reaction depends only on the initial and final states of the system but not on the reaction pathway.
- ☐ Hess's Law and enthalpy diagrams can be used for calculating $\Delta H$ values for chemical and physical changes.
- ☐ The symbol ° refers to standard conditions, where all participating species are in their standard states at $p = 100$ kPa and a specified temperature (typically 298 K).
- ☐ The standard enthalpy change ($\Delta H^\circ$) of a process is the enthalpy change of that process under standard conditions.
- ☐ The standard enthalpy of formation ($\Delta H_f^\circ$) of a substance is the standard enthalpy change upon formation of 1 mol of that substance from elementary substances.
- ☐ $\Delta H^\circ(\text{reaction}) = \Sigma\Delta H_f^\circ(\text{products}) - \Sigma\Delta H_f^\circ(\text{reactants})$
- ☐ The standard enthalpy of combustion ($\Delta H_c^\circ$) of a substance is the enthalpy change of the reaction where 1 mol of the substance is completely burnt in oxygen under standard conditions.
- ☐ $\Delta H^\circ(\text{reaction}) = \Sigma\Delta H_c^\circ(\text{reactants}) - \Sigma\Delta H_c^\circ(\text{products})$
- ☐ The thermal effects of chemical and physical processes can be determined experimentally using a calorimeter or estimated from bond enthalpies (BE).
- ☐ $\Delta H^\circ(\text{reaction}) = \Sigma\text{BE}(\text{reactants}) - \Sigma\text{BE}(\text{products})$
- ☐ Entropy ($S$) is a measure of the disorder of a system.
- ☐ Under the same conditions, gases and dissolved species tend to have greater entropies than solids and liquids.
- ☐ An increase in entropy ($\Delta S^\circ > 0$) and decrease in enthalpy ($\Delta H^\circ < 0$) are independent factors that both favour spontaneous physical and chemical changes.
- ☐ Average reaction rate ($v_{avr}$) is the change in concentration ($\Delta c$) of a reactant or product per unit time ($\Delta t$).
- ☐ Overall reaction rate is equal to the reaction rate with respect to a particular species divided by the stoichiometric coefficient of that species.
- ☐ Reaction rates can be determined experimentally by measuring changes in pressure, volume or mass of the reaction mixture or a particular species.
- ☐ Instantaneous reaction rate ($v_{inst}$) is the change in concentration ($dc$) of a reactant or product over an infinitesimally small period of time ($dt$).
- ☐ Initial reaction rate ($v_{init}$) is the instantaneous reaction rate at $t = 0$.
- ☐ Instantaneous reaction rate can be determined by plotting the concentration of a reactant or product against time and drawing a tangent line to the curve.
- ☐ Reaction rates depend on concentrations of reactants, pressure (for gases), temperature and the presence of catalysts.
- ☐ Average kinetic energies ($E_{kin}$) of reacting species increase with temperature.
- ☐ Activation energy ($E_a$) is the minimum energy required for a reaction to occur.
- ☐ Chemical changes can take place only when reactant particles collide with each other in the correct orientation and possess sufficient kinetic energy ($E_{kin} \geq E_a$).
- ☐ A catalyst increases the reaction rate without being consumed in that reaction by providing an alternative reaction pathway with a lower activation energy.
- ☐ Chemical equilibrium is a dynamic state where the forward and reverse reactions occur at equal rates.
- ☐ Position of chemical equilibrium is characterized by the equilibrium constant ($K_c$), which is the ratio of the equilibrium concentrations of reactants and products raised to the power of their stoichiometric coefficients.
- ☐ Reaction quotient ($Q_c$) is calculated in the same way as $K_c$ but using actual concentrations of reacting species instead of their equilibrium concentrations.

- ☐ If $Q_c < K_c$, the forward reaction is dominant; if $Q_c > K_c$, the reverse reaction is dominant; if $Q_c = K_c$, the system is at equilibrium.
- ☐ According to Le Châtelier's principle, a system at equilibrium counteracts any changes in reaction conditions by shifting the equilibrium position.
- ☐ Changes in concentration, pressure or volume may affect the equilibrium position but not the $K_c$ value, which depends only on temperature and the identity of reacting species.
- ☐ A catalyst increases the rates of forward and reverse reactions equally, so it has no effect on the equilibrium position or the $K_c$ value.

## Additional problems

1. A mixture of potassium chlorate and magnesium metal is used in fireworks. When the mixture is ignited, a highly exothermic reaction occurs:

$$KClO_3(s) + 3Mg(s) \rightarrow KCl(s) + 3MgO(s)$$

   Calculate the standard enthalpy change for this reaction using the following data:

$$2KClO_3(s) \rightarrow 2KCl(s) + 3O_2(g) \qquad \Delta H^\circ = -77.6 \text{ kJ}$$

$$2Mg(s) + O_2(g) \rightarrow 2MgO(s) \qquad \Delta H^\circ = -1203.6 \text{ kJ}$$

2. Methylamine, $CH_3NH_2$, is an organic base. At SATP, methylamine is a colourless gas with a strong fishy odour.
   a) State the equations that represent $\Delta H^\circ_f$ and $\Delta H^\circ_c$ for this compound.
   b) Estimate $\Delta H^\circ_c$ for methylamine using bond enthalpies from table 4.
   c) Calculate $\Delta H^\circ_f$ for methylamine using your answer to part (b) and the data from table 1.
   d) According to literature, $\Delta H^\circ_f(CH_3NH_2) = -22.5 \text{ kJ mol}^{-1}$. Suggest why your answer to part (c) differs from that value.

3. Suggest which of the following processes can be studied using a coffee-cup calorimeter:
   a) $H_2O(s) \rightarrow H_2O(l)$
   b) $H_2O(g) \rightarrow H_2O(l)$
   c) $2CO(g) + O_2(g) \rightarrow 2CO_2(g)$
   d) $AgNO_3(aq) + NaCl(aq) \rightarrow AgCl(s) + NaNO_3(aq)$

4. Describe a possible experimental procedure that can be used for determining the standard enthalpy change for one of the processes from the previous problem.

5. Using Hess's Law and the experimental values of $\Delta H^\circ_n$ from worked example 3 and question 8, calculate the standard enthalpy change for the following process:

$$CH_3COOH(aq) \rightleftharpoons CH_3COO^-(aq) + H^+(aq)$$

6. Define, in your own words, the following terms: a) average reaction rate; b) instantaneous reaction rate; c) initial reaction rate; d) activation energy; e) catalyst.

7. Consider the following process:

$$3I^-(aq) + H_2O_2(aq) + 2H^+(aq) \rightarrow I_3^-(aq) + 2H_2O(l)$$

   Over a period of 2.0 min, the concentration of iodide ions in the solution decreased from 0.050 to 0.020 mol dm$^{-3}$. Calculate:
   a) average reaction rates, in mol dm$^{-3}$ s$^{-1}$, with respect to $I^-(aq)$ and $H_2O_2(aq)$;
   b) overall average reaction rate expressed in the same units.

8. Sketch a graph to show the changes in concentrations of $I^-(aq)$, $H_2O_2(aq)$ and $I_3^-(aq)$ over time using the data from the previous problem. Assume that the initial concentrations of $I^-(aq)$ and $H_2O_2(aq)$ were equal to each other, and the initial concentration of $I_3^-(aq)$ was zero.

9. Calcium oxide reacts with carbon dioxide as follows:

$$CaO(s) + CO_2(g) \rightarrow CaCO_3(s)$$

Excess calcium oxide was placed into a pressurized vessel with a volume of 1.00 dm³. The vessel was filled with gaseous carbon dioxide and sealed off. A constant temperature of 25.0°C was maintained during the whole experiment. The plot of the pressure inside the vessel against time is shown in figure 20.

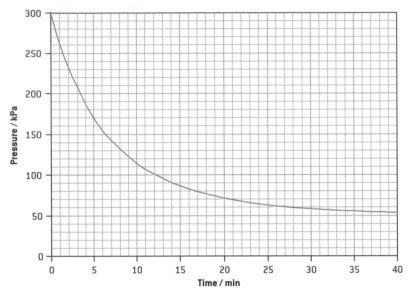

**Figure 20**

a) Using the ideal gas law (*1.3 Stoichiometric relationships*), calculate the concentrations of $CO_2(g)$ in the vessel at $t$ = 0, 5, 10, ..., 40 min. Assume that the volume occupied by solid substances is negligible.

b) Calculate the average reaction rate, in mol dm⁻³ min⁻¹, for the period between 0 and 20 min.

c) Plot the concentration of carbon dioxide against time on graph paper.

d) Using the tangent line method, determine the initial reaction rate and the instantaneous reaction rate at $t$ = 10 min.

10. Consider the effects of various factors on the reaction rate and complete the table below. In each case, assume that all other reaction conditions remain unchanged. Some cells are already filled as examples.

| Factor | Frequency of collisions | Average $E_{kin}$ of reacting species | Activation energy of reaction | Rate of forward reaction |
|---|---|---|---|---|
| decrease in a reactant concentration | | | | |
| increase in pressure of a gaseous reactant | increases | | | |
| decrease in pressure of a gaseous product | | no change | | |
| increase in volume of a gaseous reaction mixture | | | no change | |
| decrease in temperature | | | | decreases |
| increase in surface area of a solid reactant | | | | |
| addition of a catalyst | | | | |

11. State the $K_c$ expressions for the following equations:
    a) $3O_2(g) \rightleftharpoons 2O_3(g)$
    b) $2NO_2(g) \rightleftharpoons 2NO(g) + O_2(g)$
    c) $NO_2(g) \rightleftharpoons NO(g) + 0.5O_2(g)$
    d) $CH_3COOH(aq) \rightleftharpoons CH_3COO^-(aq) + H^+(aq)$

12. At a certain temperature, the $K_c$ value for reaction (b) from additional problem 11 above is 0.81. Deduce the $K_c$ value for reaction (c) from the same problem.

13. Hydrogen iodide was heated in a sealed vessel at 1100 K until the following equilibrium was reached:

    $$2HI(g) \rightleftharpoons H_2(g) + I_2(g)$$

    a) Calculate the $K_c$ value for this equation if the equilibrium concentrations of HI(g), $H_2$(g) and $I_2$(g) were 0.20, 0.12 and 0.12 mol dm$^{-3}$, respectively.
    b) Determine the initial concentration of gaseous hydrogen iodide.
    c) The $K_c$ value for the same equation at 1000 K is 0.32. Deduce the direction of the spontaneous process in the reaction mixture from (a) when it is cooled down from 1100 to 1000 K.
    d) Compare the $K_c$ values at 1000 and 1100 K and deduce the sign of the standard enthalpy change for this reaction.

14. Consider the following equilibrium:

    $$2CO(g) + O_2(g) \rightleftharpoons 2CO_2(g) \quad \Delta H° = -566 \text{ kJ}$$

    Outline how the following changes will affect the equilibrium position and the $K_c$ value of this reaction:
    a) decrease in pressure;
    b) increase in temperature;
    c) increase in concentration of oxygen gas;
    d) decrease in concentration of carbon monoxide;
    e) addition of a catalyst.

15. At high temperature, ammonia decomposes into nitrogen and hydrogen:

    $$2NH_3(g) \rightleftharpoons N_2(g) + 3H_2(g)$$

    The progress of this reaction under certain conditions without a catalyst is shown in Figure 21.

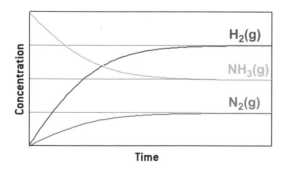

**Figure 21**

On the same graph, sketch the changes in concentrations of $NH_3$(g), $N_2$(g) and $H_2$(g) over time for the same reaction under the same conditions in the presence of a catalyst.

# 6 Acids and bases

> " In a very real sense, we can make an acid be anything we wish — the differences between the various acid–base concepts are not concerned with which is 'right' but which is most convenient to use in a particular situation. "
>
> **James E. Huheey, *Inorganic Chemistry: Principles of Structure and Reactivity* (1978)**

## Chapter context

In the previous chapters, you learned that **acids** and **bases** are two classes of chemical compounds with opposing properties. One of the most common reaction types, **neutralization**, usually involves an acid and a base as reactants and a **salt** and **water** as products. At the same time, some salts and even water itself demonstrate acidic or basic properties by reacting with bases or acids, respectively. Therefore, we need to take a closer look at the **chemical nature** of acids and bases, their **classification** and their **behaviour** in **aqueous solutions**.

## Learning objectives

In this chapter you will learn about:

→ how to define acids and bases using **acid–base theories**

→ classification, properties and **chemical equilibria** of acids and bases

→ the **concept of pH**

→ determining and **predicting acidity** of solutions in the laboratory

 **Key terms introduced**

→ Arrhenius acids and bases

→ Brønsted acid and bases

→ Dissociation and ionization

→ Conjugate acid–base pairs

→ Monoprotic and diprotic acids and bases; monobasic and dibasic acids; monoacidic and diacidic bases

→ Hydronium (hydroxonium) ion

→ Amphoteric and amphiprotic species

→ Acid dissociation constant, $K_a$

→ Ionic product of water, $K_w$

→ The pH scale

→ Acid–base indicators

→ Buffer solutions

 **DP link**

The IB Chemistry Diploma Programme covers this entire topic in **8 Acids and bases**.

 **Practical skills: Safe laboratory practices**

In the past, chemists often identified acids, bases and other compounds by their smell, taste and feel. This practice could lead to serious injury or even death. You should never taste or touch any laboratory chemicals, and avoid inhaling their vapours by carrying out experiments in the fume cupboard.

## 6.1 Acid–base theories

Acids and bases have been known for thousands of years. The term "acid" is derived from the Latin word *acere*, which means "sour" and refers to the characteristic taste of vinegar, lemon juice and other acidic solutions. Basic substances, such as potash (potassium carbonate) and lime water (a solution of calcium hydroxide) were used by ancient Egyptians for making soap and parchment. People who worked with basic solutions noted their slippery, soap-like feeling to the touch, bitter taste and ability to react with acids.

The opposing nature of acids and bases was not fully recognized until the 17th century. The first rational approach to these compounds was proposed by Robert Boyle, an Irish natural philosopher and a pioneer of modern science. According to Boyle, acids and bases can be defined as follows:

* *acids* taste sour, react with metals, turn litmus red, and can be neutralized by bases

* *bases* feel slippery, turn litmus blue, and can be neutralized by acids.

**Scientific theories and empirical rules**

Boyle's definitions of acids and bases emphasize the most characteristic properties of these compounds (reactivity towards metals and each other) and suggest a simple experimental procedure (the colour change of litmus, a natural acid–base indicator) for distinguishing between acidic and basic solutions. This approach is a good illustration of the scientific method, which is based on systematic observations and experimental evidence.

At the same time, Boyle was unable to explain why some compounds behaved as acids and others as bases. In the 17th century, the chemical composition of most substances was still unknown, and even the existence of chemical elements was not universally accepted. As a result, Boyle's classification had no theoretical background or predictive power, so it was not a scientific theory but rather a set of empirical rules.

**Figure 1.** Robert William Boyle (1627–1691)

The first scientific theory of acids was proposed in the second half of the 18th century by the French chemist Antoine Lavoisier, who worked with his wife Marie-Anne. According to his theory, all acids contained oxygen, and the strength of the acid increased with the number oxygen atoms in its molecule. Indeed, Lavoisier's theory correctly described the composition of all acids known at that time and could explain why, for example, sulfuric acid, $H_2SO_4$, containing four oxygen atoms in the molecule was stronger than sulfurous acid, $H_2SO_3$, with only three oxygen atoms. However, this theory was soon disproved by the discovery that hydrogen chloride, HCl, hydrogen sulfide, $H_2S$, and other similar compounds contained no oxygen but still behaved as typical acids.

**Figure 2.** Antoine Lavoisier (1743–1794) and Marie-Anne Lavoisier (1758–1836)

Lavoisier was also the first to find out that sulfur was an individual element rather than a compound. That discovery permitted the determination of the chemical composition of hydrogen sulfide and led to the downfall of the oxygen theory of acids. Like many other scientific theories, Lavoisier's hypothesis was falsified by experimental evidence and eventually replaced by the understanding that hydrogen, not oxygen, is an essential element of any acid.

## The Arrhenius theory

In 1884, the Swedish scientist Svante Arrhenius formulated the modern hydrogen theory of acids and defined acids and bases in terms of the ions they produced in aqueous solutions:

- *an Arrhenius acid* is a substance that dissociates in water to form hydrogen ions ($H^+$)
- *an Arrhenius base* is a substance that dissociates in water to form hydroxide ions ($OH^-$).

🔑 **Key term**

An **Arrhenius acid** dissociates in water to form hydrogen ions, while an **Arrhenius base** dissociates in water to form hydroxide ions.

### The hydronium ion

Modern studies show that a free proton, $H^+$, cannot exist in aqueous solutions, because it immediately reacts with water and produces a *hydronium ion* (also known as *hydroxonium*), $H_3O^+$:

In the above scheme, the oxygen atom donates one of its lone electron pairs to the empty orbital of the hydrogen cation. All three O–H bonds in the hydronium ion have identical lengths, and the overall shape of the ion is a trigonal pyramid.

In this book, we will be using the $H^+$ symbol as a shorthand equivalent of $H_3O^+$. However, we should always remember that all acid–base processes in aqueous solutions involve hydronium ions rather than isolated protons.

**Internal link**

The three-dimensional shapes of molecules and ions are discussed in **2.3 Chemical bonding**.

For example, hydrogen chloride is an acid, as it produces $H^+$ ions when dissolved in water:

$$HCl(aq) \rightarrow H^+(aq) + Cl^-(aq)$$

Similarly, sodium hydroxide is a base, as it produces $OH^-$ ions in aqueous solutions:

$$NaOH(aq) \rightarrow Na^+(aq) + OH^-(aq)$$

According to Arrhenius, neutralization is a combination reaction between $H^+$ and $OH^-$ ions:

$$H^+(aq) + OH^-(aq) \rightarrow H_2O(l)$$

The last reaction is the only chemical process that takes place when a strong acid is mixed with a strong base, as shown in worked example 1.

**Worked example: Ionic equations of neutralization reactions**

1. Deduce the molecular and net ionic equations for the neutralization of sulfuric acid by potassium hydroxide.

*Solution*

$$H_2SO_4(aq) + 2KOH(aq) \rightarrow K_2SO_4(aq) + 2H_2O(l)$$

$$2H^+(aq) + SO_4^{2-}(aq) + 2K^+(aq) + 2OH^-(aq) \rightarrow 2K^+(aq) + SO_4^{2-}(aq) + 2H_2O(l)$$

$$2H^+(aq) + 2OH^-(aq) \rightarrow 2H_2O(l)$$

$$H^+(aq) + OH^-(aq) \rightarrow H_2O(l)$$

Note that the net ionic equation is the same as that for the reaction between HCl(aq) and NaOH(aq) above.

**Question**

1  Deduce the molecular and net ionic equations for the neutralization of **a)** nitric acid with sodium hydroxide, **b)** hydrogen bromide with barium hydroxide.

**Figure 3.** Reaction of hydrogen chloride with ammonia

The Arrhenius theory has certain limitations. First, all Arrhenius acids and bases must be soluble in water, or they will not be able to produce any ions in aqueous solutions. Second, some bases, such as ammonia, $NH_3$, contain no oxygen and thus cannot produce hydroxide ions by dissociation. When dissolved in water, ammonia does indeed produce some $OH^-$ ions by reacting reversibly with water:

$$NH_3(aq) + H_2O(l) \rightleftharpoons NH_4^+(aq) + OH^-(aq)$$

However, the source of hydroxide ions in this reaction is water, not ammonia itself. Therefore in the absence of water, no $OH^-$ ions could be formed. Nevertheless, gaseous ammonia behaves as a typical base. For example, it readily reacts with gaseous hydrogen chloride, producing a white "smoke" of ammonium chloride (figure 3):

$$NH_3(g) + HCl(g) \rightarrow NH_4Cl(s)$$

This neutralization reaction cannot be explained by the Arrhenius theory, as it does not involve hydroxide ions.

## The Brønsted–Lowry theory

To overcome these limitations, a new acid–base theory was proposed in 1923 by two physical chemists, Johannes Brønsted from Denmark and Martin Lowry from England. Working independently of each other, Brønsted and Lowry came to the conclusion that both acids and bases can be defined by their roles in the transfer of protons ($H^+$):

- a *Brønsted acid* is a proton donor (i.e., a species that can lose an $H^+$ ion);
- a *Brønsted base* is a proton acceptor (i.e., a species that can gain an $H^+$ ion).

In the reaction with gaseous ammonia, hydrogen chloride formally loses a proton and thus acts as a Brønsted acid:

$$HCl(g) \rightarrow H^+ + Cl^-$$

In turn, ammonia accepts a proton and thus acts as a Brønsted base:

$$NH_3(g) + H^+ \rightarrow NH_4^+$$

The ammonium cation and chloride anion form ammonium chloride:

$$NH_4^+ + Cl^- \rightarrow NH_4Cl(s)$$

If we add the last three equations together, the $H^+$, $NH_4^+$ and $Cl^-$ ions will cancel one another, and the resulting equation will represent the overall neutralization reaction:

$$NH_3(g) + HCl(g) \rightarrow NH_4Cl(s)$$

According to the Brønsted–Lowry theory, the reaction of ammonia with water is also a neutralization process. Water acts as an acid by losing a proton while ammonia acts as a base by accepting a proton:

$$H_2O(l) \rightleftharpoons H^+(aq) + OH^-(aq)$$

$$NH_3(aq) + H^+(aq) \rightleftharpoons NH_4^+(aq)$$

**Figure 4.** Top: Johannes Brønsted (1879–1947); bottom: Thomas Lowry (1874–1936)

 **Key term**

A **Brønsted acid** is a proton donor.

A **Brønsted base** is a proton acceptor.

## Key term

The reaction of ammonia with water is usually referred to as ionization rather than dissociation. While these two terms are often used interchangeably, there is a subtle difference between them. **Dissociation** means that a single molecule (or other species) breaks into two or more parts, while **ionization** refers to any process that produces ions. Since the reaction of ammonia with water involves more than one species, it should not be called dissociation.

When these two equations are combined together, the resulting equation represents the ionization of ammonia in water:

$$NH_3(aq) + H_2O(l) \rightleftharpoons NH_4^+(aq) + OH^-(aq)$$

### Question

2  Deduce the equation for the reaction between hydrogen chloride and water.

It is important to note that the Brønsted–Lowry approach to acids and bases does not replace the Arrhenius theory, but rather expands it by removing any references to the solvent (water) and recognizing a wider range of species as acids and bases. For example, any Arrhenius acid, such as HCl(aq), will be also a Brønsted acid, as it acts as a proton donor. At the same time, many Arrhenius bases, such as sodium hydroxide (NaOH), are treated by the Brønsted–Lowry theory as complexes of a base ($OH^-$) with a metal cation ($Na^+$). Indeed, sodium hydroxide cannot act as a proton acceptor without losing another ion ($Na^+$), while the hydroxide ion can:

$$OH^-(aq) + H^+(aq) \rightarrow H_2O(l)$$

In this book, we will be using the term "base" in the broad sense and apply it to both NaOH and $OH^-$, as either species has an obvious basic nature.

---

### DP ready | Theory of knowledge

#### Lewis acids and bases

Every scientific theory has its limitations. For example, many oxides of nonmetals, such as $CO_2$ or $SO_3$, behave as typical acids in reactions with bases, of example:

$$CO_2(g) + NaOH(aq) \rightarrow NaHCO_3(aq)$$
$$CO_2(g) + OH^-(aq) \rightarrow HCO_3^-(aq)$$

However, neither Arrhenius nor Brønsted–Lowry theory recognizes such oxides as acids, as they lack hydrogen. The acid–base behaviour of hydrogen-free species was first explained by the American physical chemist Gilbert Lewis, who defined acids as acceptors of electron pairs and bases as donors of electron pairs. In one of his works, Lewis wrote that "restricting the group of acids to those substances that contain hydrogen interferes as seriously with the systematic understanding of chemistry as would the restriction of the term oxidizing agent to substances containing oxygen".

The Lewis theory of acids and bases is one of the cornerstones of modern organic chemistry, which is discussed in more detail in the IB Chemistry AHL Diploma Programme.

---

An interesting consequence of the Brønsted–Lowry theory is that the same species can behave both as an acid and a base. For example, water can lose a proton and produce a hydroxide anion, or accept a proton and produce a hydronium cation:

$$H_2O(l) \rightleftharpoons H^+(aq) + OH^-(aq) \text{ (water acts as an acid)}$$
$$H_2O(l) + H^+(aq) \rightleftharpoons H_3O^+(aq) \text{ (water acts as a base)}$$

Moreover, one molecule of water can pass a proton to another molecule of water, in which case the first molecule will act as an acid and the second molecule as a base:

$$H_2O(l) + H_2O(l) \rightleftharpoons H_3O^+(aq) + OH^-(aq)$$

Water and other species that can be both Brønsted acids and Brønsted bases are often called *amphiprotic* (as they can accept or donate a proton). A broader term, *amphoteric*, refers to species that can react with both acids and bases (and thus have both acidic and basic properties). Any amphiprotic species is amphoteric by definition (if it can donate a proton, it is an acid, and if it can accept a proton, it is a base). However, not all amphoteric species are amphiprotic. For example, zinc oxide can react with both acids and bases, so it is amphoteric:

$$ZnO + 2HCl \rightarrow ZnCl_2 + H_2O$$

$$ZnO + 2NaOH \rightarrow Na_2ZnO_2 + H_2O$$

At the same time, ZnO cannot donate a proton (as it has none), so it is not amphiprotic.

>  **Key term**
>
> **Amphiprotic** species can accept or donate a proton.
>
> **Amphoteric** species can react with both acids and bases.

### Question

3   Consider the following species: HF, $F^-$, $NH_3$, $NH_4^+$, $H_3PO_4$, $H_2PO_4^-$, $HPO_4^{2-}$, $PO_4^{3-}$, $Al_2O_3$, $Al^{3+}$. Which of these species can act as Brønsted acids, Brønsted bases, or both? Which of them are amphiprotic, and which ones are amphoteric?

### Conjugate acid–base pairs

Another consequence of the Brønsted–Lowry theory is that any Brønsted acid that loses a proton produces a Brønsted base, and in turn, any Brønsted base that gains a proton produces a Brønsted acid. The acid–base pairs where the species differ by exactly one proton are called *conjugate acid–base pairs*. For example, hydrogen cyanide (HCN) acts as a Brønsted acid by losing a proton and producing a conjugate base, the cyanide anion ($CN^-$):

> **Key term**
>
> The species in a **conjugate acid–base pair** differ by exactly one proton.

$$HCN(aq) \rightleftharpoons H^+(aq) + CN^-(aq)$$
conjugate acid          conjugate base

If we expand this equation to include a molecule of water, two different conjugate acid–base pairs will be formed:

$$HCN(aq) + H_2O(l) \rightleftharpoons H_3O^+(aq) + CN^-(aq)$$
conjugate   conjugate   conjugate   conjugate
acid 1        base 2        acid 2        base 1

### Question

4   Methylamine, $CH_3NH_2$, is an organic base. Its ionization in aqueous solutions proceeds as follows:

$$CH_3NH_2(aq) + H_2O(l) \rightleftharpoons CH_3NH_3^+(aq) + OH^-(aq)$$

Identify two conjugate acid–base pairs in this equation and state which species act as Brønsted acids and which as Brønsted bases.

It is very important to remember that in a conjugate acid–base pair, the acid and the base differ by exactly one proton (the acid has one more proton than the base). For example, in the following equation, sulfuric acid ($H_2SO_4$) and sulfate ion ($SO_4^{2-}$) do *not* form a conjugate acid–base pair, as they differ by two protons instead of one:

$$H_2SO_4(aq) \rightarrow 2H^+(aq) + SO_4^{2-}(aq)$$

However, if we write the equations for stepwise dissociation of sulfuric acid, each of these equations will contain a pair of conjugate acid and base:

$$H_2SO_4(aq) \rightarrow H^+(aq) + HSO_4^-(aq)$$

conjugate acid        conjugate base

$$HSO_4^-(aq) \rightarrow H^+(aq) + SO_4^{2-}(aq)$$

conjugate acid        conjugate base

Note that once again, the same species ($HSO_4^-$) acts as a Brønsted base in one process and as a Brønsted acid in another.

**Question**

5  Using the table below, deduce the formulae of conjugate acids and bases for each species. The first two rows are already filled for you.

| Species | Conjugate acid | Conjugate base |
|---|---|---|
| $H_2O$ | $H_3O^+$ | $OH^-$ |
| $Cl^-$ | $HCl$ | does not exist |
| $HF$ | | |
| $NH_3$ | | |
| $(CH_3)_3N$ | | |
| $HCO_3^-$ | | |
| $CO_3^{2-}$ | | |

The Arrhenius and Brønsted–Lowry theories emphasize the difference between acidic and basic species, but tell us very little about their individual properties, such as strength and reactivity. These properties are discussed in the next section.

## 6.2 Classification and properties of acids and bases

### Acids

Nearly all common acids are covered by the Arrhenius theory discussed in the previous topic. Each acid must contain at least one *exchangeable* (weakly bound) hydrogen atom that can detach from the rest of the acid molecule. Exchangeable hydrogen atoms usually form bonds with highly electronegative atoms, such as oxygen, halogens or sulfur. In almost all *organic acids*, exchangeable hydrogen atoms are bonded to oxygen.

For example, hydrogen chloride, sulfuric and ethanoic (acetic) acids have the following structural formulae:

hydrogen chloride          sulfuric acid          ethanoic acid

The exchangeable hydrogen atoms are shown in blue. In aqueous solutions, these hydrogen atoms dissociate and form $H^+$ (or $H_3O^+$) cations, while the remaining part of the acid produces an anion:

$$HCl(aq) \rightarrow H^+(aq) + Cl^-(aq)$$

$$H_2SO_4(aq) \rightarrow 2H^+(aq) + SO_4^{2-}(aq)$$

$$CH_3COOH(aq) \rightleftharpoons H^+(aq) + CH_3COO^-(aq)$$

Note that although ethanoic acid contains four hydrogen atoms, only one of them is exchangeable. To explain this fact, we need to look at the electronegativities of hydrogen ($\chi = 2.2$), carbon ($\chi = 2.6$) and oxygen ($\chi = 3.4$). Hydrogen and carbon have similar electronegativities ($\Delta\chi = 2.6 - 2.2 = 0.4$), so the C–H bond has low polarity and does not break easily. In contrast, the difference in electronegativity between hydrogen and oxygen is significant ($3.4 - 2.2 = 1.2$), so the O–H bond is highly polar. Since the bonding electron pair is shifted towards the more electronegative O atom, the less electronegative H atom develops a partial positive charge. As a result, it dissociates readily to form an $H^+$ ion.

In inorganic acids containing oxygen (oxoacids), all hydrogen atoms are usually bonded to oxygen, and so are exchangeable. For example, both hypochlorous (HClO) and chlorous ($HClO_2$) acids contain O–H fragments rather than Cl–H:

hypochlorous acid          chlorous acid

 **DP link**

Organic acids are described in **10 Organic chemistry**.

## Question

6   Draw the structural formulae of the following oxoacids: chloric ($HClO_3$), perchloric ($HClO_4$), carbonic ($H_2CO_3$) and phosphoric ($H_3PO_4$).

Depending on the number of exchangeable hydrogen atoms, acids are classified as:

- *monoprotic* (one exchangeable hydrogen atom), for example, HCl
- *diprotic* (two exchangeable hydrogen atoms), for example, $H_2SO_4$
- *triprotic* (three exchangeable hydrogen atoms), for example, $H_3PO_4$.

In contrast to inorganic acids, organic acids often contain both exchangeable and nonexchangeable hydrogen atoms. For example, both methanoic and ethanoic acids are monoprotic, despite the fact that their molecules contain two and four hydrogen atoms, respectively.

The formulae and names of common acids and their anions are given in table 1 on page 156. Along with the systematic names, many organic acids have trivial names, which are shown in brackets.

**Key term**

**Monoprotic acids** (with one **exchangeable** hydrogen atom) are sometimes called **monobasic** (as they can react with one equivalent of a base), **diprotic acids** are called **dibasic**, and so on. Similarly, bases can be classified as **monoacidic**, **diacidic**, and so on. In this book, we will use the term "-protic" for both acids and bases.

**Internal link**

The nomenclature of organic acids will be discussed in 7.2 Classification and nomenclature of organic compounds.

Table 1. Common acids and their anions

| Acid | | | Anion | |
|---|---|---|---|---|
| Formula | Name | Strength | Formula | Name |
| HF | hydrogen fluoride | weak | $F^-$ | fluoride |
| HCl | hydrogen chloride | strong | $Cl^-$ | chloride |
| HBr | hydrogen bromide | strong | $Br^-$ | bromide |
| HI | hydrogen iodide | strong | $I^-$ | iodide |
| $H_2S$ | hydrogen sulfide | weak | $S^{2-}$ | sulfide |
| HCN | hydrogen cyanide | weak | $CN^-$ | cyanide |
| $HNO_3$ | nitric | strong | $NO_3^-$ | nitrate |
| $HNO_2$ | nitrous | weak | $NO_2^-$ | nitrite |
| $H_2SO_4$ | sulfuric | strong | $SO_4^{2-}$ | sulfate |
| $H_2SO_3$ | sulfurous | weak | $SO_3^{2-}$ | sulfite |
| $H_3PO_4$ | phosphoric | weak | $PO_4^{3-}$ | phosphate |
| $H_3PO_3$ | phosphorous | weak | $PO_3^{3-}$ | phosphite |
| $HClO_4$ | perchloric | strong | $ClO_4^-$ | perchlorate |
| $HClO_3$ | chloric | strong | $ClO_3^-$ | chlorate |
| $HClO_2$ | chlorous | weak | $ClO_2^-$ | chlorite |
| HClO | hypochlorous | weak | $ClO^-$ | hypochlorite |
| $H_2CO_3$ | carbonic | weak | $CO_3^{2-}$ | carbonate |
| HCOOH | methanoic (formic) | weak | $HCOO^-$ | methanoate (formate) |
| $CH_3COOH$ | ethanoic (acetic) | weak | $CH_3COO^-$ | ethanoate (acetate) |
| $H_2C_2O_4$ | ethanedioic (oxalic) | weak | $C_2O_4^{2-}$ | ethanedioate (oxalate) |

**Key term**

**Hydrogen chloride and hydrochloric acid**

Although the terms hydrogen chloride and hydrochloric acid refer to the same substance, HCl, they have different meanings in chemistry. When we say **hydrogen chloride**, we mean an individual compound, HCl, which is a gas under normal conditions, while **hydrochloric acid** is a solution of HCl in water. Therefore, it is incorrect to say "a solution of hydrochloric acid", as "hydrochloric acid" already refers to a solution.

Similar problems may arise when we talk about sulfuric acid, which is often used as an aqueous solution but can also exist in pure form (so called "100% sulfuric acid"). When this difference is important, we should always say "aqueous sulfuric acid" when we refer to a solution, or "anhydrous sulfuric acid" when we refer to pure $H_2SO_4$.

An important characteristic of any acid is its strength. *Strong acids*, such as hydrogen chloride, dissociate completely in aqueous solutions. If we dissolve one mole of HCl in water, the resulting solution will contain one mole of hydrogen cations and one mole of chloride anions but no HCl molecules. In other words, the dissociation of HCl is irreversible, which is represented by the single arrow:

$$HCl(aq) \rightarrow H^+(aq) + Cl^-(aq)$$

In addition to hydrogen chloride, six other strong acids are listed in table 1. You are advised to memorize their formulae and names.

Weak acids, such as ethanoic acid, dissociate only to a small extent when dissolved in water. For example, table vinegar (an aqueous solution of ethanoic acid) contains both $CH_3COOH$ molecules and the products of their dissociation, $H^+$ and $CH_3COO^-$ ions. The reversible nature of this process is represented by the equilibrium sign:

$$CH_3COOH(aq) \rightleftharpoons H^+(aq) + CH_3COO^-(aq)$$

Almost all organic and many inorganic acids are weak, so if an acid is not listed in table 1, it is safe to assume that it is weak as well. There are a few exceptions, but they are rarely discussed in general chemistry courses.

## Question

7 Write the dissociation schemes for the following acids: hydrogen bromide, hydrogen cyanide, sulfuric acid and methanoic acid. Do not forget about using a single arrow for strong acids and an equilibrium sign for weak acids.

**Internal link**

Strong and weak electrolytes, including acids and bases, are defined in **4.3 Solutions of electrolytes**.

The strength of oxoacids generally increases with the oxidation state of the central atom. In turn, a higher oxidation state usually means that the acid molecule contains more oxygen atoms. For example, the nitrogen atom in weak nitrous acid, $HNO_2$, is bound to two oxygen atoms and has an oxidation state of $+3$. An addition of the third oxygen increases the oxidation state of nitrogen from $+3$ to $+5$ and produces strong nitric acid, $HNO_3$. Similarly, strong sulfuric acid, $H_2SO_4$, has a higher oxidation state of sulfur and more oxygen atoms than weak sulfurous acid, $H_2SO_3$.

## Question

8 Write down the formulae of all oxoacids of chlorine and phosphorus from table 1. Deduce the oxidation states of these elements in each acid and state how they affect the acid strength.

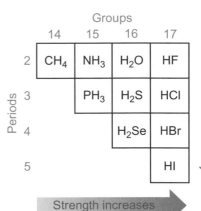

Figure 5. Periodic trends in the strength of binary acids

Binary acids demonstrate clear periodic trends: their strength increases along the period and down the group (figure 5). For example, in the third period, phosphine ($PH_3$) does not show any acidic properties in aqueous solutions, hydrogen sulfide ($H_2S$) is a weak acid, and hydrogen chloride is a strong acid. Similarly, down group 17, hydrogen fluoride (HF) is a weak acid while the other three hydrogen halides (HCl, HBr and HI) are strong acids.

### DP ready    Nature of science

**The dissociation constant**

Like any other equilibrium, the dissociation of a weak acid can be characterized by the equilibrium constant. This constant is called the *dissociation constant*, $K_a$, where the subscript index "a" means "acid". If a weak acid HA(aq) dissociates into $H^+$(aq) and $A^-$(aq) ions, its dissociation constant is expressed as follows:

$$K_a = \frac{[H^+][A^-]}{[HA]}$$

Dissociation constants provide a simple way of comparing the strengths of different acids: the higher the $K_a$, the stronger the acid.

**Internal link**

Equilibrium constants are described in **5.3 Chemical equilibrium**.

Polyprotic weak acids dissociate stepwise; for example:

$$H_2SO_3(aq) \rightleftharpoons H^+(aq) + HSO_3^-(aq)$$

$$HSO_3^-(aq) \rightleftharpoons H^+(aq) + SO_3^{2-}(aq)$$

The second proton dissociates to a much smaller extent than the first, so nearly all $H^+$ ions produced by a polyprotic acid are formed on its first dissociation step. The reason for that becomes clear if we look at the charges on the ions involved in the above equations. The first step produces $H^+$ and $HSO_3^-$ ions. These ions exert electrostatic attraction on each other, so pulling them apart requires some energy. On the second step, the electrostatic attraction between $H^+$ and $SO_3^{2-}$ ions is much greater, as the anion $SO_3^{2-}$ is doubly charged. As a result, the second step requires more energy, which makes this process less likely to occur.

## Question

9 Write out the scheme for stepwise dissociation of phosphoric acid.

## Bases

Despite the difference in chemical properties, bases show many similarities to acids in terms of their behaviour in aqueous solutions. Where an acid produces an $H^+$ ion, a base either produces an $OH^-$ ion (Arrhenius base) or accepts an $H^+$ ion (Brønsted base). However, the general principles behind these processes remain the same and thus can be explained by similar concepts and equations.

Most inorganic bases are metal hydroxides, which contain a metal atom and one or more OH groups. The nature of the chemical bond between the metal and the OH group depends on the metal electronegativity. Alkali metals (Li, Na, K, Rb and Cs) and most group 2 metals (Mg, Ca, Sr and Ba) have very low electronegativities, so they form ionic hydroxides. Such hydroxides consist of a metal ion ($M^{n+}$) and one or more hydroxide anions ($OH^-$).

All ionic hydroxides are strong bases. With the exception of $Mg(OH)_2$ and $Ca(OH)_2$, they are readily soluble in water and fully dissociate into ions, for example:

$$NaOH(aq) \rightarrow Na^+(aq) + OH^-(aq)$$

$$Ba(OH)_2(aq) \rightarrow Ba^{2+}(aq) + 2OH^-(aq)$$

Calcium hydroxide is only slightly soluble in water, and magnesium hydroxide is almost insoluble. If an excess of such hydroxide is added to water, a *heterogeneous* equilibrium between the solid base and aqueous ions is established:

$$Mg(OH)_2(s) \rightleftharpoons Mg^{2+}(aq) + 2OH^-(aq)$$

$$Ca(OH)_2(s) \rightleftharpoons Ca^{2+}(aq) + 2OH^-(aq)$$

It is important to note that these hydroxides are strong bases, so their solutions contain no undissociated molecules, $Mg(OH)_2$ and $Ca(OH)_2$. The reversible nature of the above processes is caused by low solubility of these bases in water, not by their low strength.

Less active metals, such as beryllium, aluminium and all transition elements, form covalent hydroxides, where the metal atom and the oxygen of the OH group are linked together by a polar covalent bond. For example, both $Fe(OH)_2$ and $Fe(OH)_3$ are covalent hydroxides.

Almost all covalent hydroxides are weak bases. In addition, these hydroxides are virtually insoluble in water, so they can demonstrate their basic nature only in reactions with acids.

Ammonia ($NH_3$) is one of the few inorganic bases that does not contain a metal. As you learned in the previous section, aqueous ammonia acts as a weak Brønsted base by accepting a proton from water:

$$NH_3(aq) + H_2O(l) \rightleftharpoons NH_4^+(aq) + OH^-(aq)$$

> ### Worked example: Amines as bases
>
> **2.** Organic derivatives of ammonia, *amines*, are also weak bases. In aqueous solutions, amines accept protons and produce $OH^-$ ions in the same way as ammonia. Formulate the equation for methylamine, $CH_3NH_2$, acting as a Brønsted base in aqueous solution.
>
> ***Solution***
>
> $$CH_3NH_2(aq) + H_2O(l) \rightleftharpoons CH_3NH_3^+(aq) + OH^-(aq)$$

> ### Question
>
> 10  Deduce the ionization schemes for aqueous dimethylamine and trimethylamine.

> 🔑 **Key term**
>
> **Amines** are organic derivatives of ammonia in which one or more H atoms at nitrogen are replaced with organic substituents, such as a methyl group, $CH_3$:
>
> $$CH_3NH_2$$
> methylamine
>
> $$(CH_3)_2NH$$
> dimethylamine
>
> $$(CH_3)_3N$$
> trimethylamine

Similar to acids, bases are classified as monoprotic, diprotic, triprotic, and so on:

- *monoprotic bases* can either accept a single proton (Brønsted bases) or donate a single $OH^-$ ion (Arrhenius bases). Ammonia, amines and sodium hydroxide are examples of monoprotic bases.
- *diprotic Arrhenius bases*, such as $Ca(OH)_2$, have two exchangeable OH groups
- *triprotic Arrhenius bases*, such as $Al(OH)_3$, have three exchangeable OH groups.

The anions of weak acids can also act as Brønsted bases. For example, the ethanoate ion is produced by the dissociation of ethanoic acid:

$$CH_3COOH(aq) \rightleftharpoons H^+(aq) + CH_3COO^-(aq)$$

If we reverse this equation, the basic nature of the ethanoate ion will become obvious:

$$CH_3COO^-(aq) + H^+(aq) \rightleftharpoons CH_3COOH(aq)$$

Since the ethanoate ion can accept only one proton, it is a monoprotic base. In aqueous solutions, the ethanoate ion reacts with water, producing an $OH^-$ ion:

$$CH_3COO^-(aq) + H_2O(l) \rightleftharpoons CH_3COOH(aq) + OH^-(aq)$$

Such reactions explain why the salts of weak acids and strong bases produce slightly basic solutions when dissolved in water.

Anions of polyprotic acids behave as polyprotic bases; for example:

$$SO_3^{2-}(aq) + H^+(aq) \rightleftharpoons HSO_3^-(aq)$$

$$HSO_3^-(aq) + H^+(aq) \rightleftharpoons H_2SO_3(aq)$$

These processes are similar to the stepwise dissociation of weak polyprotic acids, except that all reactions are now reversed.

11 State the equations in which the following ions act as Brønsted bases: a) cyanide ion, $CN^-$; b) phosphate ion, $PO_4^{3-}$.

## Neutralization reactions

Acids and bases tend to neutralize each other when mixed together. However, the extent of this process depends on the strengths of both the acid and the base. There are four possible cases:

1. strong acid and strong base

2. strong acid and weak base

3. weak acid and strong base

4. weak acid and weak base.

We will illustrate each case with a typical example, writing all equations in molecular, total ionic and net ionic forms. The net ionic equations will give us some idea about the nature of the neutralization process.

### 1. Neutralization of a strong acid with a strong base

This case was already discussed in *6.1 Acid–base theories*, with hydrogen chloride as a typical strong acid and sodium hydroxide as a typical strong base:

$$HCl(aq) + NaOH(aq) \rightleftharpoons NaCl(aq) + H_2O(l)$$

$$H^+(aq) + Cl^-(aq) + Na^+(aq) + OH^-(aq) \rightleftharpoons Na^+(aq) + Cl^-(aq) + H_2O(l)$$

$$H^+(aq) + OH^-(aq) \rightleftharpoons H_2O(l)$$

In the total ionic equation, the spectator ions $Na^+$ and $Cl^-$ are present on both sides, so we can cancel them out. The final solution is neutral, as it contains equal amounts of $H^+$ and $OH^-$ ions. The net ionic equation also suggests that the neutralization is almost irreversible, as water is a very weak electrolyte. Therefore, the equilibrium signs in all three equations can be replaced with single arrows.

### 2. Neutralization of a strong acid with a weak base

$$HCl(aq) + NH_3(aq) \rightleftharpoons NH_4Cl(aq)$$

$$H^+(aq) + Cl^-(aq) + NH_3(aq) \rightleftharpoons NH_4^+(aq) + Cl^-(aq)$$

$$H^+(aq) + NH_3(aq) \rightleftharpoons NH_4^+(aq)$$

In the total ionic equation, $Cl^-$ is a spectator ion. The net ionic equation shows that some $H^+$ ions are still present, so we can expect the resulting solution to be slightly acidic. This makes perfect sense: since the acid is stronger than the base, their reaction product retains some acidic nature.

### 3. Neutralization of a weak acid with a strong base

$$CH_3COOH(aq) + NaOH(aq) \rightleftharpoons CH_3COONa(aq) + H_2O(l)$$

$$CH_3COOH(aq) + Na^+(aq) + OH^-(aq) \rightleftharpoons CH_3COO^-(aq) + Na^+(aq) + H_2O(l)$$

$$CH_3COOH(aq) + OH^-(aq) \rightleftharpoons CH_3COO^-(aq) + H_2O(l)$$

In the total ionic equation, $Na^+$ is a spectator ion. Note that ethanoic acid is a weak electrolyte, so most of its molecules remain undissociated in aqueous solutions. Therefore, its formula is written as $CH_3COOH$ in all equations.

The net ionic equation shows the presence of some $OH^-$ ions, so we can expect the final solution to be slightly basic. Again, this makes sense: since the base is stronger than the acid, their reaction product retains some basic nature.

### 4. Neutralization of a weak acid with a weak base

$$CH_3COOH(aq) + NH_3(aq) \rightleftharpoons CH_3COONH_4(aq)$$

$$CH_3COOH(aq) + NH_3(aq) \rightleftharpoons CH_3COO^-(aq) + NH_4^+(aq)$$

In this case, no ions are spectators. The resulting solution is neutral, as it contains no additional $H^+$ or $OH^-$ ions (except those produced by the dissociation of water).

> **Question**
>
> 12 Write molecular and ionic equations for the neutralization reactions between **a)** sulfuric acid and ammonia; **b)** hydrogen cyanide and potassium hydroxide; **c)** hydrogen fluoride and methylamine. In each case, state the strengths of the acid and base, and predict whether the resulting solution will be neutral, acidic or basic.

Many neutralization reactions involve metal oxides and salts of weak acids and bases. For example, magnesium oxide reacts readily with hydrochloric acid to produce a salt and water:

$$MgO(s) + 2HCl(aq) \rightarrow MgCl_2(aq) + H_2O(l)$$

The nature of this process is better represented by ionic equations:

$$MgO(s) + 2H^+(aq) + 2Cl^-(aq) \rightarrow Mg^{2+}(aq) + 2Cl^-(aq) + H_2O(l)$$

$$MgO(s) + 2H^+(aq) \rightarrow Mg^{2+}(aq) + H_2O(l)$$

The net ionic equation shows that magnesium oxide accepts two $H^+$ ions and thus acts as a Brønsted base.

Metal carbonates also react with acids, producing unstable carbonic acid, $H_2CO_3$:

$$Na_2CO_3(aq) + 2HCl(aq) \rightarrow 2NaCl(aq) + H_2CO_3(aq)$$

Carbonic acid quickly decomposes into water and carbon dioxide, which bubbles out of the solution (figure 6):

$$H_2CO_3(aq) \rightarrow CO_2(g) + H_2O(l)$$

These two reactions are often written together:

$$Na_2CO_3(aq) + 2HCl(aq) \rightarrow 2NaCl(aq) + CO_2(g) + H_2O(l)$$

**Figure 6.** Reaction of baking soda with an acid

Again, the ionic equations reveal the nature of this process:

$$2Na^+(aq) + CO_3^{2-}(aq) + 2H^+(aq) + 2Cl^-(aq) \rightarrow 2Na^+(aq) + 2Cl^-(aq) + CO_2(g) + H_2O(l)$$

$$CO_3^{2-}(aq) + 2H^+(aq) \rightarrow CO_2(g) + H_2O(l)$$

As in the previous example, the anion of weak carbonic acid, $CO_3^{2-}$, acts as a Brønsted base by accepting two protons.

Metal hydrogencarbonates, such as baking soda ($NaHCO_3$), react with acids in the same way as carbonates:

$$NaHCO_3(aq) + HCl(aq) \rightarrow NaCl(aq) + CO_2(g) + H_2O(l)$$

$$Na^+(aq) + HCO_3^-(aq) + H^+(aq) + Cl^-(aq) \rightarrow Na^+(aq) + Cl^-(aq) + CO_2(g) + H_2O(l)$$

$$HCO_3^-(aq) + H^+(aq) \rightarrow CO_2(g) + H_2O(l)$$

In this case, the hydrogencarbonate ion, $HCO_3^-$, acts as a Brønsted base by accepting a proton.

> **Question**  Ⓠ
>
> 13 Write molecular and ionic equations for the reaction of sulfuric acid with **a)** calcium oxide; **b)** calcium carbonate; **c)** potassium hydrogencarbonate.

**Figure 7.** "Milk of magnesia", a suspension of magnesium hydroxide in water, is a common antacid

**DP link**

Antacids are discussed in **D.4 pH regulation of the stomach**.

**Internal link**

Redox reactions are discussed in **3.3 Redox processes**.

---

**DP ready** | **Nature of science**

**Antacids**

Heartburn and other symptoms of indigestion are caused by excess hydrochloric acid in the stomach. These symptoms can be alleviated by medicines known as antacids. The active ingredients in antacids are weak bases, such as metal oxides, hydroxides, carbonates, and hydrogencarbonates. All these compounds neutralize the excess acid, as described in the main text.

Like any pharmaceutical drugs, antacids have various side effects. In particular, carbon dioxide produced in the body from the reaction of stomach acid with carbonates and hydrogencarbonates causes bloating and belching, while the intake of metal ions disturbs the electrolyte balance in the body.

---

Before we move to the next topic, note that the reactions of acids with metals cannot be classified as acid–base or neutralization reactions. In contrast to bases, metals do not accept $H^+$ ions but reduce them to molecular hydrogen, for example:

$$Mg(s) + 2HCl(aq) \rightarrow MgCl_2(aq) + H_2(g)$$

$$Mg(s) + 2H^+(aq) \rightarrow Mg^{2+}(aq) + H_2(g)$$

Such processes are classified as redox reactions.

## 6.3 The concept of pH

So far, we have been discussing acids and bases separately by emphasizing their opposing chemical nature. However, the Brønsted–Lowry theory provides us with an important link between acids and bases, as they both participate in the transfer of $H^+$ ions. In this topic, we are going to take one step further and show that the acidity or basicity of a solution can be characterized by a single value, known as the *potential of hydrogen*, or simply pH.

## Ionic product of water

As you already know, water is a weak electrolyte that dissociates into $H^+$ and $OH^-$ ions:

$$H_2O(l) \rightleftharpoons H^+(aq) + OH^-(aq)$$

Like any other equilibrium, the dissociation of water can be characterized by the equilibrium constant ($K_c$):

$$K_c = \frac{[H^+][OH^-]}{[H_2O]}$$

where $[H^+]$, $[OH^-]$ and $[H_2O]$ are the equilibrium concentrations of participating species. As the density of pure water at room temperature is approximately 1000 g dm⁻³, the mass of each dm³ of water is about 1000 g, and the amount of water in 1 dm³ is:

$$m(H_2O)/M(H_2O) = 1000 \text{ g}/18.02 \text{ g mol}^{-1} \approx 55.5 \text{ mol}$$

Since water is a very weak electrolyte, nearly all its molecules exist in undissociated form. Therefore, in any dilute solution, the equilibrium concentration of water will have approximately the same value, 55.5 mol dm⁻³. Now we can transform the $K_c$ equation as follows:

$$K_c[H_2O] = [H^+][OH^-]$$

Both factors on the left, $K_c$ and $[H_2O]$, are constants, so their product is also a constant. This new constant, $K_w = K_c[H_2O]$, is called the *ionic product of water*:

$$K_w = [H^+][OH^-]$$

At room temperature (25°C), $K_w = 1.00 \times 10^{-14}$. This value is very small, so water is a very weak electrolyte.

The value of $K_w = 1.00 \times 10^{-14}$ can be used only for dilute aqueous solutions. In a concentrated solution (for example, in a battery acid containing 30% $H_2SO_4$), the concentration of water becomes significantly less than 55.5 mol dm⁻³, which affects the $K_w$ and thus the product of $[H^+]$ and $[OH^-]$. In addition, the ionic product of water increases with temperature. In this book, we will assume that all solutions are dilute and have a temperature of 25°C, so the $K_w$ value remains constant.

In pure water, $[H^+] = [OH^-]$. Indeed, if we remove a proton from water, a hydroxide ion will be left, so the amounts (and thus the concentrations) of these ions in pure water will always be equal to each other. If $[H^+] = [OH^-] = x$ mol dm⁻³, then $K_w = x^2$, and so each concentration can be found as a square root of $K_w$:

$$[H^+] = [OH^-] = \sqrt{1.00 \times 10^{-14}} = 1.00 \times 10^{-7} \text{ mol dm}^{-3}$$

This means that in each 55.5 mol of water, only $1.00 \times 10^{-7}$ mol exists as ions. In other words, only one in approximately 2,000,000,000 water molecules dissociates into $H^+$ and $OH^-$ ions while the rest of the molecules remain undissociated. This is the reason why in all ionic equations water is represented as $H_2O$ molecules rather than $H^+$ and $OH^-$ ions.

Many electrolytes, such as acids and bases, affect the concentrations of $H^+$ and $OH^-$ ions in aqueous solutions. For example, if we dissolve hydrogen chloride in water, more $H^+$ ions will be produced:

$$HCl(aq) \rightarrow H^+(aq) + Cl^-(aq)$$

In accordance with Le Châtelier's principle, the increased concentration of $H^+$ ions will shift the position of equilibrium of the water dissociation to the left:

$$H_2O(l) \rightleftharpoons H^+(aq) + OH^-(aq)$$

As a result, the concentration of $OH^-$ ions will decrease. Therefore, in an acidic solution, $[H^+]$ will always be greater than $[OH^-]$.

**Internal link**

Le Châtelier's principle is described in **5.3 Chemical equilibrium**.

**Worked example: Using $K_w$** WE

**3.** Calculate the concentration of $OH^-$ ions in a 0.100 mol dm$^{-3}$ solution of hydrogen chloride at 25°C.

*Solution*

A 0.100 mol dm$^{-3}$ HCl solution will contain 0.100 mol dm$^{-3}$ $H^+$ ions:

$$HCl(aq) \rightarrow H^+(aq) + Cl^-(aq)$$

| $c$, mol dm$^{-3}$ | 0.100 | 0.100 | 0.100 |

Since $K_w = [H^+][OH^-] = 1.00 \times 10^{-14}$ mol dm$^{-3}$, we can find the concentration of $OH^-$ ions as follows:

$$1.00 \times 10^{-14} = 0.100 \times [OH^-]$$

$$[OH^-] = 1.00 \times 10^{-14}/0.100 = 1.00 \times 10^{-13} \text{ mol dm}^{-3}$$

As expected, $0.100 > 1.00 \times 10^{-13}$, so $[H^+] > [OH^-]$

An addition of a base will increase the concentration of $OH^-$ ions in the solution and decrease the concentration of $H^+$ ions. Therefore, in a basic solution, $[H^+]$ will be lower than $[OH^-]$.

**Question** Q

14 Calculate the concentrations of $H^+$ and $OH^-$ ions in a 0.0500 mol dm$^{-3}$ aqueous solution of potassium hydroxide.

**The pH scale**

The values of $K_w$, $[H^+]$ and $[OH^-]$ in aqueous solutions can be very small, so they take a lot of space and are awkward to work with. Such small quantities are often represented by their negative decimal logarithms, also known as p-numbers:

$$pK_w = -\log K_w$$

$$pH = -\log[H^+]$$

$$pOH = -\log[OH^-]$$

For pure water at 25°C:

- $K_w = 1.00 \times 10^{-14}$
- $[H^+] = [OH^-] = 1.00 \times 10^{-7}$ mol dm$^{-3}$.

In this case:

- $pK_w = -\log(1 \times 10^{-14}) = 14$
- $pH = pOH = -\log(1 \times 10^{-7}) = 7$.

**Maths skills**

Working with logarithms requires some practice, but greatly simplifies the calculations. The expression "$pK_w = 14$" is more compact than "$K_w = 1.00 \times 10^{-14}$" and less likely to cause errors when written. Also, with logarithms and p-numbers, we can use addition and subtraction instead of multiplication and division:

$$\log(a \times b) = \log a + \log b$$

$$\log(a/b) = \log a - \log b$$

As a result, formulae with logarithms or p-numbers are easier to memorize and use in calculations. For example, some expressions from the main text can be transformed as shown in table 2.

**Table 2.** Useful expressions involving $K_w$, $[H^+]$ and $[OH^-]$

| Without p-numbers | With p-numbers |
|---|---|
| $[H^+][OH^-] = 1 \times 10^{-14}$ | $pH + pOH = 14$ |
| $[H^+] = \dfrac{1 \times 10^{-14}}{[OH^-]}$ | $pH = 14 - pOH$ |
| $[OH^-] = \dfrac{1 \times 10^{-14}}{[H^+]}$ | $pOH = 14 - pH$ |

The use of pH values offers an easy way of comparing the acidity and basicity of aqueous solutions over a broad range of $H^+$ and $OH^-$ concentrations (table 3). Indeed, since pH and pOH are related by a simple expression, we need to know only one of these parameters to fully characterize the solution.

**Table 3.** The pH scale

| $[H^+]$ | pH | Indicator colour | Example |
|---|---|---|---|
| $1 \times 10^{-14}$ | 14 | violet | liquid drain cleaner |
| $1 \times 10^{-13}$ | 13 | violet | bleach |
| $1 \times 10^{-12}$ | 12 | purple | ammonia solution |
| $1 \times 10^{-11}$ | 11 | purple | mild detergent |
| $1 \times 10^{-10}$ | 10 | blue | toothpaste |
| $1 \times 10^{-9}$ | 9 | blue | baking soda |
| $1 \times 10^{-8}$ | 8 | green | seawater |
| $1 \times 10^{-7}$ | 7 | green | pure water |
| $1 \times 10^{-6}$ | 6 | yellow | urine |
| $1 \times 10^{-5}$ | 5 | yellow | black coffee |
| $1 \times 10^{-4}$ | 4 | orange | tomato juice |
| 0.001 | 3 | orange | orange juice |
| 0.01 | 2 | red | vinegar |
| 0.1 | 1 | red | stomach acid |
| 1 | 0 | dark red | battery acid |

 **Maths skills**

Acidic solutions with high concentration of $H^+$ ions have low pH, while basic solutions with low $[H^+]$ have high pH. This follows from the definition $pH = -\log[H^+]$: the higher the p-number, the lower its negative logarithm. Similarly, a low pOH means high $[OH^-]$, and high pOH means low $[OH^-]$.

Pure water and solutions with pH = 7 are called neutral, as they have equal concentrations of $H^+$ and $OH^-$ ions. The pH of an acidic solution is less than 7, while the pH of a basic solution is greater than 7. These definitions are summarized in table 4.

**Table 4.** Acidic, neutral and basic aqueous solutions

| Solution | pH | pOH | [H⁺] and [OH⁻] |
|----------|-----|-----|------------------|
| acidic | < 7 | > 7 | $[H^+] > [OH^-]$ |
| neutral | 7 | 7 | $[H^+] = [OH^-]$ |
| basic | > 7 | < 7 | $[H^+] < [OH^-]$ |

Earlier in this topic we found that a 0.100 mol dm⁻³ solution of hydrogen chloride contains 0.100 mol dm⁻³ H⁺ ions. The pH value of such solution will be –log 0.100 = 1.00. As expected, 1.00 < 7, so the solution is acidic.

### Question

15 Calculate the pH values for the following aqueous solutions:
   a) 0.0100 mol dm⁻³ sulfuric acid;
   b) 0.0100 mol dm⁻³ potassium hydroxide.

### Practical skills

In a modern chemical laboratory, the pH of a solution can be measured with high accuracy and precision by a digital pH meter (figure 8). Alternatively, the pH can be estimated using *universal indicator*, which gradually changes colour across the whole pH range from 0 to 14 (table 3). Natural acid–base indicators can be obtained from plants.

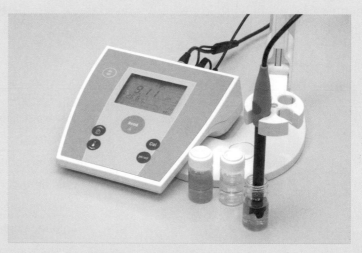

**Figure 8.** Digital pH meter

### Internal link

Titration is described in 4.2 Concentration expressions and stoichiometry.

In many cases, the exact pH of a solution is not important, and it is sufficient to know whether the solution is acidic, neutral or basic. Such situations are common when the concentrations of acids or bases are determined by *titration*. In a typical titration experiment, one of the reactants (acid or base) is slowly added to another until the neutralization is complete. At this moment, the pH of the solution changes sharply, which can be detected by an acid–base indicator. Universal indicator is not suitable for titrations, as its colour changes gradually over a broad pH range. Therefore, other indicators with two contrasting colours are commonly used (table 5).

**Table 5.** Common acid–base indicators

| Indicator | Colour | | |
|---|---|---|---|
| | **Acidic (low pH)** | **Neutral (pH = 7)** | **Basic (high pH)** |
| Methyl orange | red | yellow | yellow |
| Litmus | red | purple | blue |
| Phenolphthalein | colourless | colourless | fuchsia |

In other experiments, it is important to maintain a constant pH even when small amounts of acids or bases are added to the solution. This can be achieved by the use of acid–base *buffers*, the solutions that contain both components of a conjugate acid–base pair.

Let's consider a mixture of sodium hydrogencarbonate and sodium carbonate, containing the following ions:

$$NaHCO_3(aq) \rightarrow Na^+(aq) + HCO_3^-(aq)$$

$$Na_2CO_3(aq) \rightarrow 2Na^+(aq) + CO_3^{2-}(aq)$$

The hydrogencarbonate and carbonate anions form a conjugate acid–base pair. They do not react with each other but can neutralize other acids or bases. If a strong acid is added to this solution, it will react with the base, $CO_3^{2-}$, and produce the conjugate acid, $HCO_3^-$:

$$CO_3^{2-}(aq) + H^+(aq) \rightleftharpoons HCO_3^-(aq)$$

If a strong base is added, it will react with the acid, $HCO_3^-$, and produce the conjugate base, $CO_3^{2-}$:

$$HCO_3^-(aq) + OH^-(aq) \rightleftharpoons CO_3^{2-}(aq) + H_2O(l)$$

In both cases, the concentrations of $HCO_3^-$ and $CO_3^{2-}$ ions will change slightly, but the overall composition of the solution will remain the same. Therefore, the pH of the solution will not change significantly.

**Question**

16 An acid–base buffer contains ethanoic acid and sodium ethanoate. Identify the conjugate acid–base pair of this buffer and write the reactions that demonstrate its buffer action towards strong acids and bases.

 **Key term**

**Buffers** are solutions containing a conjugate acid–base pair. They are able to resist changes in pH when strong acids or bases are added to the solution.

**DP link**

Acid–base buffers are very important components of all living organisms. For example, the pH of human blood is kept within a very narrow range of 7.35–7.45 units by several buffer systems involving hydrogencarbonate and hydrogenphosphate ions, carbon dioxide, proteins and other substances. You can learn more about these topics in **B Biochemistry** and **D Medicinal Chemistry**.

## Chapter summary

We have now discussed all the topics required for the understanding of acid–base processes in aqueous solutions. Before moving further, please make sure that you have a working knowledge of the following concepts and definitions:

☐ An Arrhenius acid produces protons ($H^+$ ions) in aqueous solutions, and an Arrhenius base produces $OH^-$ ions.

☐ A Brønsted acid is a proton donor, and a Brønsted base is a proton acceptor.

☐ An acid forms its conjugate base by losing a proton, and a base forms its conjugate acid by accepting a proton.

- ☐ Amphiprotic species, such as water, can both donate and accept protons.
- ☐ Net ionic equations reveal the chemical nature of acid–base processes.
- ☐ Strong acids and bases are fully dissociated (ionized) in aqueous solutions while weak acids and bases undergo only partial dissociation (ionization).
- ☐ In pure water and neutral solutions $[H^+]$ = $[OH^-]$.
- ☐ In acidic solutions $[H^+]$ > $[OH^-]$, and in basic solutions $[H^+]$ < $[OH^-]$.
- ☐ At 25°C, the ionic product of water ($K_w$) is $1 \times 10^{-14}$, and $pK_w$ = 14.
- ☐ pH = $-\log[H^+]$.
- ☐ At 25°C, a solution is neutral at pH = 7, acidic at pH < 7, and basic at pH > 7.
- ☐ The pH of a solution can be measured by a digital pH meter or estimated using acid–base indicators.
- ☐ Buffer solutions contain conjugate acid–base pairs that maintain constant pH by neutralizing small amounts of strong acids and bases.

## Additional problems

1. In 1810, the British chemist Sir Humphry Davy determined the elemental composition of hydrogen sulfide and thus disproved Lavoisier's oxygen theory of acids. In one of his papers, Davy concluded that "acidity does not depend upon any particular elementary composition, but upon peculiar arrangement of various substances". Discuss whether this statement was a scientific theory or not.

2. Compare and contrast the Arrhenius and Brønsted–Lowry theories of acids and bases.

3. A scheme on page 150 in *6.1 Acid–base theories* shows a proton accepting a lone electron pair from a molecule of water and producing a hydronium ion. Draw a similar scheme for the formation of an ammonium ion.

4. Deduce the molecular and net ionic equations for the neutralization of barium hydroxide with a) hydrogen chloride, b) sulfuric acid. Note that barium sulfate is insoluble in water.

5. The complete neutralization of 1.0 mol of aqueous hydrogen chloride with sodium hydroxide produces 57 kJ of heat. Predict how much heat will be released in the reaction between the same amounts of aqueous hydrogen bromide and potassium hydroxide.

6. Deduce the equations for the stepwise dissociation of hydrogen sulfide. Identify all conjugate acid–base pairs in these equations.

7. Phenylamine, $C_6H_5NH_2$, is a weak organic base. Draw the ionization scheme for this base in the aqueous solution and identify two conjugate acid–base pairs in this equation.

8. Hypophosphorous acid, $H_3PO_2$, is a rare example of an inorganic oxoacid containing nonexchangeable hydrogen atoms. In aqueous solutions, it behaves as a monoprotic acid. Deduce the structural formula of this acid and identify exchangeable and nonexchangeable hydrogen atoms in its molecule.

9. Arsenic forms several oxoacids, including arsenous ($H_3AsO_3$) and arsenic ($H_3AsO_4$). Deduce the oxidation state of arsenic in both acids and predict which acid will be the strongest.

10. Write molecular and ionic equations for the neutralization reactions between a) hydrogen sulfide and sodium hydroxide; b) perchloric acid and ammonia. In each case, state the strengths of the acid and base, and predict whether the resulting solution will be neutral, acidic or basic.

11. The ionic product of water increases with temperature. At 10°C, $K_w$ = $2.95 \times 10^{-15}$, and at 80°C, $K_w$ = $2.45 \times 10^{-13}$. a) Calculate the pH of pure water at 10 and 80°C. b) Discuss whether these pH values imply that cold water is basic and hot water is acidic.

12. Calculate the concentrations of $H^+$ and $OH^-$ ions in the following solutions:
    a) 0.0400 mol dm$^{-3}$ hydrochloric acid; b) 0.0100 mol dm$^{-3}$ barium hydroxide.

13. Calculate the pH of each solution from the previous problem.

14. Overproduction of hydrochloric acid in the stomach causes indigestion, which can be relieved by antacids. One tablet of a certain brand of antacid contains 840 mg of magnesium carbonate. a) Write the reaction between magnesium carbonate and hydrochloric acid. b) Calculate the amount (number of moles) of magnesium carbonate in one tablet of the antacid. c) Calculate the amount of hydrochloric acid that will react with magnesium carbonate. d) Estimate the volume of gastric juice with $c(HCl) = 0.040$ mol dm$^{-3}$ that will be neutralized by one antacid tablet.

15. Deduce the colours of methyl orange and phenolphthalein when each of these acid–base indicators is added to the following solutions: a) HCl; b) $Na_2CO_3$; c) $NH_4Cl$; d) NaBr; e) KOH.

16. An acid–base buffer contains potassium hydrogenphosphate, $K_2HPO_4$, and potassium dihydrogenphosphate, $KH_2PO_4$. Identify the conjugate acid–base pair of this buffer and deduce the equations that demonstrate its buffer action towards strong acids and bases.

# Organic chemistry

> "The synthesis of substances occurring in Nature, perhaps in greater measure than activities in any other area of organic chemistry, provides a measure of the conditions and powers of science. "
>
> **Robert Burns Woodward,** *Perspectives in Organic Chemistry* (1956)

## Chapter context

Among all chemical elements, **carbon** has the greatest capacity for bonding with itself, producing **chains**, **rings** and other structures of almost infinite size and complexity. At the same time, carbon atoms readily form **covalent bonds** with other elements, such as **hydrogen**, **oxygen**, **nitrogen** and **halogens**. These properties make the chemistry of carbon so rich and diverse that it is treated as a separate discipline — **organic chemistry**. Nearly all substances containing carbon are classified as **organic compounds**, as many of them are commonly found in living organisms. In contrast, elementary carbon, its oxides and some other derivatives, such as carbonic acid, carbonates, carbides and cyanides, are often produced in nature by non-biological processes and thus are considered inorganic.

## Learning objectives

In this chapter you will learn about:

→ Different classes of organic compounds

→ Their nomenclature, structures and properties.

*(before studying this material, review concepts of valence and covalent bonding, electronegativity, Lewis structures and VSEPR theory—see internal link below)*

##  Key terms introduced

→ Full and condensed structural formulae

→ Skeletal formulae

→ Functional groups

→ Homologous series

→ Structural isomers

→ Locants

→ Saturated and unsaturated hydrocarbons

→ Alkanes, alkenes and alkynes

→ Arenes (aromatic hydrocarbons): benzene, toluene and xylene

→ Complete and incomplete combustion

→ Chain reactions

→ Polymers, macromolecules and monomers

→ Distillation and reflux

→ Carbonyl compounds

→ Condensation reactions

##  Internal link

For valence and covalent bonding see **1.2 Chemical substances, formulae and equations**, for electronegativity see **2.2 The periodic law** and for Lewis structures and VSEPR theory see **2.3 Chemical bonding.**

## DP link

**10 Organic Chemistry**

## 7.1 Structures of organic molecules

Carbon is a group 14 element with four electrons in the outer energy level. Because of its moderate electronegativity ($\chi$ = 2.6), carbon rarely forms ionic bonds. Instead, it tends to share its valence electrons with other atoms and form covalent compounds.

In the ground state, the electron configuration of carbon is [He]$2s^2 2p^2$. However, this element can also exist in the excited state [He]$2s^1 2p^3$, where each valence electron occupies a separate atomic orbital (figure 1).

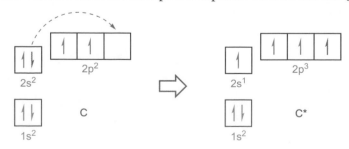

**Figure 1.** Electron configurations of the carbon atom in the ground state (left) and excited state (right)

Each electron in the excited carbon atom can pair up with an electron from another atom, producing one covalent bond. To achieve a stable octet configuration, carbon must share all four of its electrons. As a result, carbon is tetravalent in nearly all organic compounds (table 1).

**Table 1.** Structures of simple organic molecules

| Name | Molecular formula | Lewis structure |
|---|---|---|
| methane | $CH_4$ | |
| ethane | $C_2H_6$ | |
| ethene | $C_2H_4$ | |
| ethyne | $C_2H_2$ | |
| chloromethane | $CH_3Cl$ | |
| methoxymethane (dimethyl ether) | $C_2H_6O$ | |
| methanal | $CH_2O$ | |
| methanamine (methylamine) | $CH_5N$ | |
| ethanenitrile | $C_2H_3N$ | |

In the simplest organic molecule, methane ($CH_4$), each of the four C–H bonds is formed by an electron pair shared between the carbon and hydrogen atoms. In larger molecules, two carbon atoms can share one, two or three electron pairs with each other, producing single, double or triple bonds, respectively. Three or more carbon atoms can link together in different ways, producing chains and rings. Elements other than hydrogen and carbon, collectively known as *heteroatoms*, can also form single or multiple bonds with carbon atoms. The most common heteroatoms in organic molecules are oxygen, nitrogen and halogens, followed by phosphorus and sulfur.

1 Each element in an organic compound tends to form a specific number of covalent bonds and to have a specific number of lone electron pairs. Explore table 1 and complete the table below. The first row is already filled as an example.

| Element | Number of bonds | Number of lone pairs |
|---|---|---|
| C | 4 | 0 |
| H | | |
| Cl (and other halogens) | | |
| O | | |
| N | | |

2 Draw the Lewis structures for the following organic molecules: **a)** propane, $C_3H_8$; **b)** trichloromethane (chloroform), $CHCl_3$; **c)** methanol, $CH_3OH$; **d)** methanoic acid, HCOOH. Note that oxygen atoms in methanoic acid do not form covalent bonds with each other.

 **Key term**

- **Full structural formulae** show shared electron pairs as lines but omit lone electron pairs.

- In **condensed structural formulae** single bonds are omitted and monovalent atoms are grouped together.

- **Skeletal formulae** show covalent bonds but omit atom labels for all carbon and most hydrogen atoms.

The Lewis structures shown in table 1 provide detailed information about the number and distribution of valence electrons in each molecule. However, these structures take up a lot of space and require extra effort for drawing lone electron pairs. Therefore, organic molecules are often represented by simplified formulae with varying levels of detail.

*Full structural formulae* are similar to Lewis structures. In such formulae, all shared electron pairs are shown as lines and all lone pairs are omitted. For example, the full structural formula of methylamine looks as follows:

$$H-\underset{\underset{H}{|}}{\overset{\overset{H}{|}}{C}}-\underset{\diagdown H}{\overset{\diagup H}{N}}$$

In *condensed structural formulae*, all single bonds are omitted and all monovalent substituents at each atom are grouped together. For example, the condensed formula of methylamine is $CH_3NH_2$. Although double and triple bonds in condensed formulae can also be omitted, it may lead to ambiguity, so you are strongly advised to show such bonds explicitly. For example, the condensed formula of ethene should be written as $H_2C=CH_2$ or $CH_2=CH_2$, not as $CH_2CH_2$.

*Skeletal formulae* show covalent bonds as lines but omit atom labels for all carbon and some hydrogen atoms. The positions of these atoms in a skeletal formula are defined as follows:

- the end of each line or the meeting point of two or more lines symbolizes a carbon atom;

- hydrogen atoms connected to carbon atoms are not shown at all, as their number can be implied using the octet rule;

- heteroatoms, together with any attached hydrogen atoms, are shown as single groups (such as OH or $NH_2$);

- lone electron pairs on heteroatoms are usually omitted but can be shown if required.

As an example, let's consider the following skeletal formula:

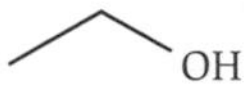

The two lines in this formula represent two covalent bonds. The left terminal and the meeting point (vertex) denote carbon atoms. At present, the terminal carbon has only one covalent bond, so it needs three hydrogen atoms to satisfy the octet rule. Therefore, the terminal group is $CH_3$. The second carbon atom has two covalent bonds, so it must form another two bonds with hydrogen atoms. Therefore, the group in the middle is $CH_2$. Now we can expand the skeletal formula to a full structural formula:

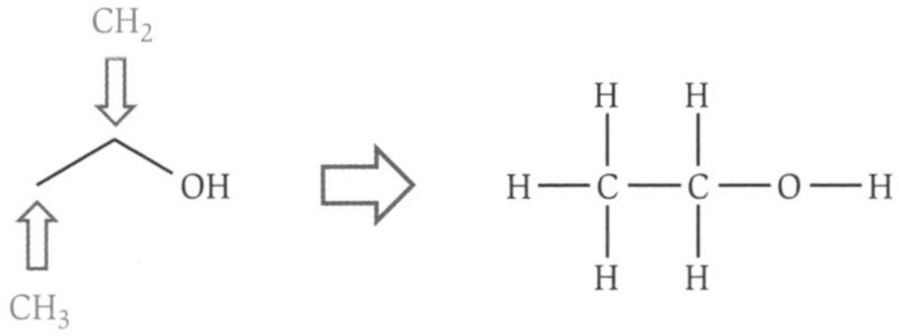

More examples of full, condensed and skeletal formulae are given in table 2.

**Table 2.** Full structural, condensed structural and skeletal formulae of selected organic molecules

| Name | Full structural formula | Condensed structural formula | Skeletal formula |
|---|---|---|---|
| propane | 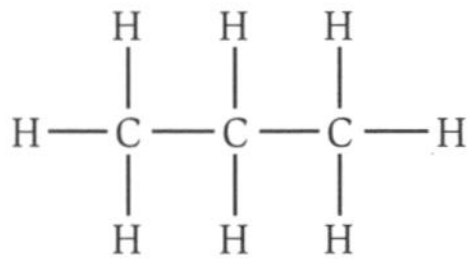 | $CH_3CH_2CH_3$ |  |
| propene | 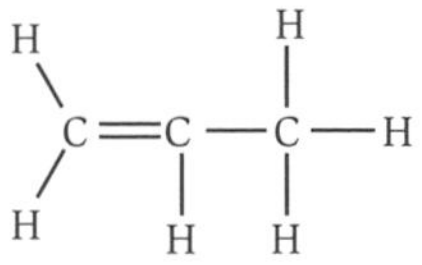 | $H_2C=CHCH_3$ | 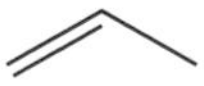 |
| 1-chloropropane | 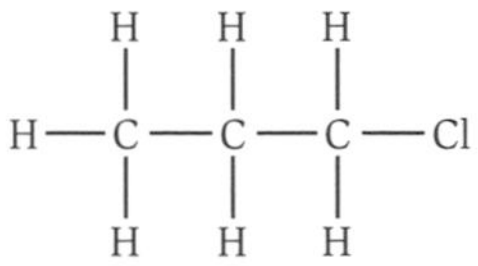 | $CH_3CH_2CH_2Cl$ | 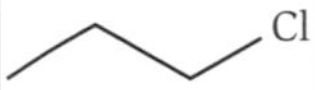 |
| propan-2-ol | 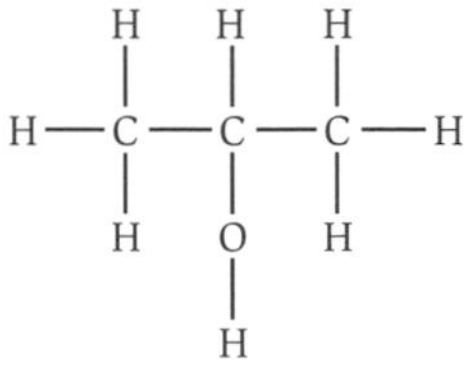 | $CH_3CH(OH)CH_3$ | 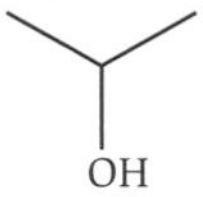 |
| propanoic acid | 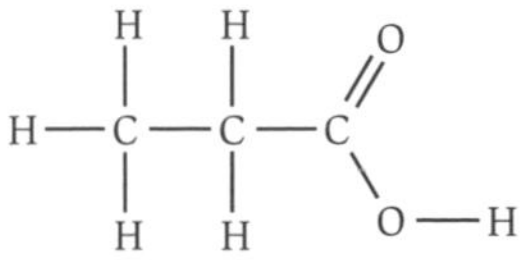 | $CH_3CH_2COOH$ | 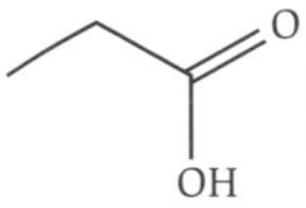 |

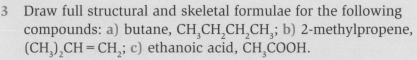
Structures of organic molecules can also be represented by so called "semi-condensed" formulae, which are intermediate between full and condensed. Such formulae show all atoms and most covalent bonds, so they are commonly used in chemistry textbooks and reference materials, including the latest version of the IB Chemistry Data Booklet. For example, semi-condensed formulae of propan-2-ol and propanoic acid from table 2 are as follows:

Other variants of semi-condensed formulae are used for cyclic compounds, which will be discussed later in this chapter.

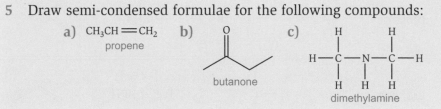
Structural formulae are drawn in two dimensions on paper, so they tell us little about three-dimensional shapes of organic molecules. These shapes can be visualized with various 3D models, where atoms and covalent bonds are shown as balls, sticks or their combinations (figure 2). Each kind of model emphasizes a particular feature of the molecular structure. For example, *stick models* are useful for showing bond angles while *space-filling models* give a more realistic picture of the distribution of electron density around individual atoms. *Ball-and-stick models* are the most common, as they offer a good compromise between representing the bond angles and the overall shapes of organic molecules. To distinguish between carbon, hydrogen and heteroatoms in molecular models, these elements are normally represented by different colours: black for carbon, white for hydrogen, red for oxygen, blue for nitrogen and green for chlorine.

**Figure 2.** Three-dimensional models of methane: stick (top), ball-and-stick (middle) and space-filling (bottom)

According to VSEPR theory, a carbon atom with four single bonds adopts a tetrahedral geometry with bond angles of approximately 109.5°. Similarly, a carbon atom with one double and two single bonds will have a trigonal planar shape with all bond angles of about 120°. Finally, one triple and one single bond at a carbon atom will form a straight line with a bond angle of 180°. Therefore, we can predict the shapes of simple organic molecules from their structural formulae (figure 3).

**Internal link**

VSEPR theory enables us to predict the bond angles around an atom — see **2.3 Chemical bonding**.

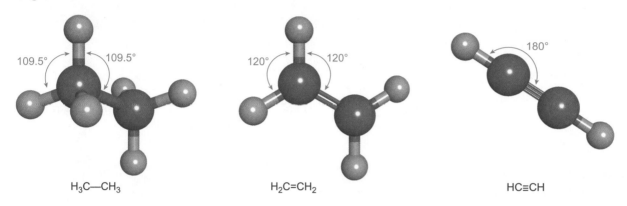

$H_3C$—$CH_3$   $H_2C$=$CH_2$   HC≡CH

**Figure 3.** Ball-and-stick models of ethane (left), ethene (middle) and ethyne (right)

Bond angles around heteroatoms, such as oxygen and nitrogen, are similar to those around carbon and can vary from 105 to 120°. Halogens in organic molecules are usually monovalent, so they do not form any specific bond angles.

**Practical skills**

Working with models is an important skill for any chemistry student, so you are advised to borrow or purchase a molecular model kit and spend some time assembling and examining the models of organic molecules mentioned in this chapter. Alternatively, you can visualize molecular structures using free online resources, such as molview.org or chemagic.org. These activities will help you to bridge the gap between structural formulae and three-dimensional shapes of molecules and speed up your progress in learning organic chemistry.

## 7.2 Classification and nomenclature of organic compounds

Despite their enormous number and variety, organic compounds can be divided into a relatively small number of *classes* containing specific arrangements of atoms, known as *functional groups*. For example, methanol ($CH_3OH$) belongs to the class of alcohols, as it contains a hydroxyl group (–OH) covalently bound to a carbon atom. Similarly, methylamine ($CH_3NH_2$) is an amine, as it has an amino group (–$NH_2$).

**Key term**

**Functional groups** are specific arrangements of atoms within molecules that are responsible for the reactions those molecules undergo. Organic compounds can be divided into **classes** according to their functional groups.

The names of many classes and functional groups are defined by the International Union of Pure and Applied Chemistry (IUPAC) and form the basis of a systematic nomenclature of organic compounds, known as the *IUPAC nomenclature*.

## Alkanes, alkenes and alkynes

The simplest organic compounds, *alkanes*, have no functional groups and consist only of carbon and hydrogen atoms bonded to each other by single covalent bonds (table 3). Alkanes belong to a larger group of organic compounds, *hydrocarbons*.

**Table 3.** The homologous series of alkanes

| IUPAC name | Molecular formula | Condensed structural formula | Ball-and-stick model |
|---|---|---|---|
| methane | $CH_4$ | $CH_4$ | |
| ethane | $C_2H_6$ | $CH_3CH_3$ | |
| propane | $C_3H_8$ | $CH_3CH_2CH_3$ | |
| butane | $C_4H_{10}$ | $CH_3CH_2CH_2CH_3$ | |
| pentane | $C_5H_{12}$ | $CH_3(CH_2)_3CH_3$ | |
| hexane | $C_6H_{14}$ | $CH_3(CH_2)_4CH_3$ | |

**Table 4.** Greek numerical prefixes

| Number | Prefix | Number | Prefix |
|---|---|---|---|
| 1 | mono- | 6 | hex(a)- |
| 2 | di- | 7 | hept(a)- |
| 3 | tri- | 8 | oct(a)- |
| 4 | tetr(a)- | 9 | non(a)- |
| 5 | pent(a)- | 10 | dec(a)- |

## Question

6 Deduce the IUPAC names of straight-chain alkanes with eight and ten carbon atoms.

All members of the alkane family have the same general formula ($C_nH_{2n+2}$; $n = 1, 2, \ldots$) and differ from each other by one or more $CH_2$ groups. Such families of compounds are known as *homologous series*. Organic compounds of other classes form their own homologous series. For example, the two most common alcohols, methanol ($CH_3OH$) and ethanol ($CH_3CH_2OH$) are homologues, as they differ from each other by a $CH_2$ group.

In addition to straight-chain molecules like those shown in table 3, the carbon backbone in alkanes can branch in one or more directions. For example, five carbon atoms can be linked together in several ways (figure 4).

**Key term**

**Homologous series of compounds** have the same general formula and differ from each other by one or more $CH_2$ groups.

**Structural isomers** have the same molecular formula but different bonding order and thus different structural formulae.

$$H_3C-CH_2-CH_2-CH_2-CH_3 \qquad \underset{\underset{CH_3}{|}}{H_3C-CH-CH_2-CH_3} \qquad \overset{\overset{CH_3}{|}}{\underset{\underset{CH_3}{|}}{H_3C-C-CH_3}}$$

**Figure 4.** Structural isomers of pentane

The above alkanes have the same molecular formula ($C_5H_{12}$) but different bonding order and thus different structural formulae. Such compounds are called *structural isomers*.

## Question

7 Deduce semi-condensed formulae of all structural isomers with the molecular formula $C_6H_{14}$.

The IUPAC name of a branched molecule is derived from the name of its longest carbon chain, also known as the *principal chain*. In figure 4, the molecule in the middle is an alkane with four carbon atoms in the longest chain, so its name will include the stem "butane". Each branch, or *side chain*, is treated as a *substituent* (as it substitutes a hydrogen atom in the longest chain) and named after the alkane with the same number of carbon atoms by changing the suffix "-ane" to "-yl" (table 5). In our case, the substituent ($CH_3–$) has a single carbon atom, so its name is "methyl". Finally, the principal chain is numbered from one end to the other in such a way as to give the lowest possible number (*locant*) to the side chain.

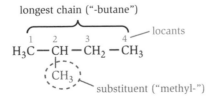

Now we can construct the systematic name of this alkane by adding the substituent name (methyl), together with its locant (2), as a prefix to the name of the longest chain (butane):

2 + methyl + butane = 2-methylbutane

Note that the locant is separated from the substituent name by a hyphen while the rest of the name is written as a single word.

**Table 5.** Common substituents in organic molecules

| Formula | Name |
|---|---|
| $CH_3-$ | methyl |
| $CH_3CH_2-$ or $C_2H_5-$ | ethyl |
| $CH_3CH_2CH_2-$ | propyl |
| $(CH_3)_2CH-$ | isopropyl |
| ⬡— or $C_6H_5-$ | phenyl |
| $-F$ | fluoro |
| $-Cl$ | chloro |
| $-Br$ | bromo |
| $-I$ | iodo |
| $-NO_2$ | nitro |

When two or more identical substituents are present in the same molecule, their locants and names are grouped together using Greek numerical prefixes from table 4:

2,2-dimethylpropane    2,2,4-trimethylpentane

## Question

8  Deduce IUPAC names for all isomers from question 7.

The names of different substituents are arranged in alphabetical order, regardless of their nature and position in the principal chain:

2-bromo-3-ethyl-2-methylpentane

### Key term

In IUPAC names, the position of each substituent is normally specified by a **locant**. However, some molecules have only one possible position for a substituent, so the locant becomes redundant and should be omitted. For example, we say "chloroethane" rather than "1-chloroethane", as no other chloroethane can exist. An attempt to place a single substituent at another carbon atom in ethane will produce the same molecule, as the principal chain must be numbered from the end closest to the substituent:

chloroethane
(not "1-chloroethane")

chloroethane
(not "1-chloroethane")

In other molecules, a substituent can be introduced at various positions, but only one type of substitution will leave the stem of the IUPAC name unchanged. For example, replacing a hydrogen atom in propane with a methyl group will produce either butane or 2-methylpropane:

butane
(not "1-methylpropane")

2-methylpropane
or methylpropane

butane
(not "3-methylpropane")

Therefore, the only possible methylpropane is 2-methylpropane, so the locant (2) can be either included in its name or omitted. There are no strict rules about such situations, so you are advised to keep redundant locants unless instructed otherwise by your teacher.

When two or more different substituents are present in the same molecule, their locants must never be omitted. For example, hydrogen atoms in ethane can be substituted with chlorine and bromine in two different ways, producing two isomeric compounds:

1-bromo-2-chloroethane

1-bromo-1-chloroethane

The only exception from this rule is methane, which contains a single carbon atom. Any substituents in methane are always quoted without their locants. For example, $CF_2Cl_2$ is called dichlorodifluoromethane, not "1,1-dichloro-1,1-difluoromethane".

 **Internal link**

Dichlorodifluoromethane is a common ozone-depleting refrigerant; see **5.2 Chemical kinetics.**

If we know the systematic name of an organic compound, we can always deduce the structural formula of that compound.

**Worked example: Constructing structural formulae of organic compounds** **WE**

**1.** Deduce the structural formula of 3-chloro-2,2-dimethylbutane.

*Solution*

The stem ("butane") suggests that this name refers to an alkane (suffix "-ane") with four carbon atoms in the principal chain (stem "but"). Therefore, the backbone of this compound will look as follows:

$$-\overset{1}{C}-\overset{2}{C}-\overset{3}{C}-\overset{4}{C}-$$

Now we can add the substituents (a chlorine atom at C-3 and two methyl groups at C-2):

To complete the structural formula, we need to add hydrogen atoms to all free valences:

3-chloro-2,2-dimethylbutane

**Question** Q

9   Draw the structural formulae for the following compounds: **a)** 2-methylhexane; **b)** 1,1,2,2-tetrachloroethane; **c)** 2,3-dibromo-3-ethylpentane.

*Cycloalkanes* are another family of hydrocarbons with three or more carbon atoms joined into a ring by single covalent bonds. These compounds with the general formula $C_nH_{2n}$ ($n = 3, 4, \ldots$) are named by adding the prefix "cyclo-" to the name of straight-chain alkanes with the same number of carbon atoms:

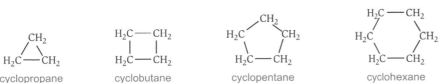

cyclopropane          cyclobutane          cyclopentane          cyclohexane

## Key term

**Saturated hydrocarbons** are compounds in which all the carbon–carbon bonds are single.

**Unsaturated hydrocarbons** are compounds containing double and triple carbon-carbon bonds.

Alkanes and cycloalkanes are collectively known as *saturated* hydrocarbons, as their molecules cannot accept any more hydrogen atoms without breaking a chain of carbon atoms. All other hydrocarbons contain multiple carbon–carbon bonds and thus are *unsaturated*, as they can be hydrogenated ("saturated" with hydrogen) while keeping their carbon skeleton intact. For example, ethene and ethyne react with hydrogen as follows:

$$H_2C=CH_2(g) + H_2(g) \rightarrow H_3C-CH_3(g) \quad HC\equiv CH(g) + 2H_2(g) \rightarrow H_3C-CH_3(g)$$

ethene           ethane         ethyne           ethane

Ethene is the first member of the homologous series of *alkenes* (table 6). Due to the presence of a carbon–carbon double bond (C=C) in their molecules, these unsaturated hydrocarbons have two hydrogen atoms fewer than alkanes with the same carbon backbone, so their general formula is $C_nH_{2n}$ ($n = 2, 3, \ldots$).

**Table 6.** The homologous series of alkenes and alkynes

| $n$ | Alkene ($C_nH_{2n}$) | | Alkyne ($C_nH_{2n-2}$) | |
|---|---|---|---|---|
| | Name | Condensed formula | Name | Condensed formula |
| 2 | ethene | $H_2C=CH_2$ | ethyne | $HC\equiv CH$ |
| 3 | propene | $H_2C=CHCH_3$ | propyne | $HC\equiv CCH_3$ |
| 4 | but-1-ene | $H_2C=CHCH_2CH_3$ | but-1-yne | $HC\equiv CCH_2CH_3$ |
| 5 | pent-1-ene | $H_2C=CH(CH_2)_2CH_3$ | pent-1-yne | $HC\equiv C(CH_2)_2CH_3$ |
| 6 | hex-1-ene | $H_2C=CH(CH_2)_3CH_3$ | hex-1-yne | $HC\equiv C(CH_2)_3CH_3$ |

Systematic names of alkenes are derived from the names of alkanes by replacing the suffix "-ane" with "-ene". The primary chain of an alkene must include the C=C bond and be numbered from the end closest to that bond. The position of the C=C bond in the primary chain is specified by its lowest locant:

$$\overset{1}{H_3C}-\overset{2}{CH}=\overset{3}{CH}-\overset{4}{CH_2}-\overset{5}{CH_3}$$

pent-2-ene

$$\overset{4}{H_3C}-\overset{3}{\underset{\underset{CH_3}{|}}{CH}}-\overset{2}{CH}=\overset{1}{CH_2}$$

3-methylbut-1-ene

$$\overset{1}{H_2C}=\overset{2}{\underset{\underset{H_2C-CH_3}{|}}{C}}-\overset{3}{CH_2}-\overset{4}{CH_3}$$

2-ethylbut-1-ene

## Key term

**Alkenes** are unsaturated hydrocarbons with a carbon–carbon double bond (C=C) and the general formula $C_nH_{2n}$ ($n = 2, 3, \ldots$).

**Alkynes** are unsaturated hydrocarbons with a carbon–carbon triple bond (C≡C) and the general formula $C_nH_{2n-2}$.

### Question

10 Deduce structural formulae and systematic names for all structural isomers with the molecular formula $C_4H_8$.

*Alkynes* are another class of unsaturated hydrocarbons (table 6). Their molecules contain a carbon–carbon triple bond (C≡C) and they have the general formula $C_nH_{2n-2}$. The names of alkynes are constructed in the same way as those of alkenes, except that the suffix "-ene" is replaced with "-yne".

### Aromatic hydrocarbons

*Arenes*, or *aromatic hydrocarbons*, contain a ring of several (typically six) carbon atoms with delocalized electrons.

The simplest arene is *benzene*, $C_6H_6$, which is sometimes represented by a structural formula with alternating single and double bonds:

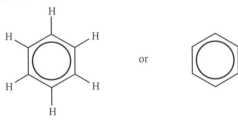

However, this structure does not explain the experimentally observed facts. All carbon–carbon bonds in benzene have equal lengths and strengths, and their enthalpies (507 kJ mol⁻¹) are intermediate between the enthalpies of C–C and C=C bonds (346 and 614 kJ mol⁻¹; see table 4 in *5.1 Thermochemistry*). To explain these facts, we need to have a closer look at the electron structure of benzene.

Each carbon atom in benzene forms three localized single bonds by sharing three of its electrons with immediate neighbours (one hydrogen and two carbon atoms). The fourth electron is donated to a common electron cloud that surrounds all carbon atoms in the aromatic ring. This cloud of six delocalized electrons (one from each carbon atom) increases the order of each carbon–carbon bond from 1 to 1.5. The structure of benzene is thus better represented by a six-membered ring with a circle inside, which symbolizes the delocalized electrons and emphasizes their symmetrical distribution within the molecule:

**Internal link**

Delocalized electrons are introduced in **2.3 Chemical bonding**.

or

**Key term**

**Arenes**, or **aromatic hydrocarbons**, contain at least one ring of several (typically six) carbon atoms with delocalized electrons.

The **benzene** homologous series has the general formula $C_nH_{2n-6}$.

Benzene and other aromatic hydrocarbons form a homologous series of arenes with the general formula $C_nH_{2n-6}$. The systematic names of benzene homologues are constructed by adding substituent prefixes to the stem "benzene":

methylbenzene
(toluene)

1,2-dimethylbenzene
(xylenes)

1,3-dimethylbenzene

**Key term**

Methylbenzene and dimethylbenzene are also known by their common names **toluene** and **xylene**, respectively. You should be able to recognize these names, as they are still used in chemical literature.

**Question**

11 Two structural isomers of xylene are shown in the text. Deduce structural formulae and systematic names for two other structural isomers of xylene that contain a six-membered aromatic ring.

In larger molecules, the benzene ring itself may act as a substituent. In such cases, it is named *phenyl* and designated as $-C_6H_5$ in condensed structural formulae (table 5).

## Organic compounds containing oxygen and nitrogen

Organic compounds often contain heteroatoms such as oxygen and nitrogen. Certain combinations of these heteroatoms are treated as functional groups, some of which are listed in table 7. When a molecule has two or more different functional groups, its class name is defined by the most senior group (higher up in the table).

**Table 7.** Common functional groups in decreasing order of priority (carbon atoms shown in black are included in the principal chain; non-hydrogen substituents are designated as R)

| Functional group | | Suffix | Prefix | Class name | Example |
|---|---|---|---|---|---|
| Formula | Name | | | | |
| | carboxyl | -oic acid | —* | carboxylic acid | ethanoic acid |
| | ester | -oate | —* | ester | methyl ethanoate |
| | amido | -amide | —* | amide | ethanamide |
| —C≡N | cyano | -nitrile | cyano- | nitrile | H₃C—C≡N ethanenitrile |
| | carbonyl | -al | oxo- | aldehyde | ethanal |
| | | -one | | ketone | propanone |
| —OH | hydroxyl | -ol | hydroxy- | alcohol | H₃C—OH methanol |
| —NH₂ | amino | -amine | amino- | amine | H₃C—NH₂ methanamine (methylamine**) |
| R—O—R′ | ether | — | -oxy- | ether | H₃C—O—CH₃ methoxymethane |

* Not required for the IB Chemistry Diploma Programme

** Alternative name accepted by IUPAC

The IUPAC rules state that the name of any organic compound is derived from the name of its simplest hydrocarbon analogue with the same number of carbon atoms. For example, the simplest analogue of methanol, methanal and methanoic acid is methane, so the names of all four compounds share the same stem, "methan":

methane    methanol    methanal    methanoic acid

Some functional groups, such as carboxyl (–COOH), can only be present at the end of a carbon chain, so their locants in IUPAC names are omitted. Other functional groups, such as hydroxyl (–OH), may appear at any carbon atom, so their positions must be specified by locants:

propanoic acid    propan-1-ol    propan-2-ol

Note that the primary carbon chain is numbered in such a way as to give the functional group the lowest possible number.

When two or more different functional groups are present in the same molecule, only the most senior group (higher up in table 7) forms a suffix while all other groups are treated as substituents and form prefixes:

4-methyl-2-oxopentanal

Note that all substituents are listed in alphabetical order, regardless of their nature. The primary carbon chain is numbered from the most senior functional group.

## Question

12 Deduce the IUPAC names for the following compounds. Refer to table 7 if needed.

The structural formula of a functionalized organic compound can be deduced from its IUPAC name using the same procedure as that described earlier for hydrocarbons.

## Worked example: Constructing structural formulae of functionalized organic compounds

**2.** Construct the formula of the common amino acid phenylalanine, which has the IUPAC name 2-amino-3-phenylpropanoic acid.

*Solution*

First of all, we analyse the name and identify the stem, suffixes and prefixes:

stem
(chain length and saturation)

2-amino-3-phenyl**propan**oic acid

prefixes and locants
(substituents and
their positions)

last suffix
(functional group)

The stem "propan" means that the primary chain consists of three carbon atoms (root "prop") and is saturated (suffix "an"):

$$-\overset{|}{\underset{|}{C}}-\overset{|}{\underset{|}{C}}-\overset{|}{\underset{|}{C}}-$$

The compound contains a carboxyl group (suffix "oic acid"), which can only exist at the end of the primary chain.

$$-\overset{|}{\underset{|}{\overset{3}{C}}}-\overset{|}{\underset{|}{\overset{2}{C}}}-\overset{1}{C}\overset{\displaystyle O}{\underset{OH}{\diagup\diagdown}}$$

The two prefixes show that there are two substituents in this molecule:

- an amino group ($-NH_2$) at the second carbon atom (prefix "2-amino")
- a phenyl group ($-C_6H_5$) at the third carbon atom (prefix "3-phenyl").

To produce the final structure, we need to add hydrogen atoms to all free valences:

2-amino-3-phenylpropanoic acid

$C_6H_5-CH_2-CH-COOH$
          |
         $NH_2$

(semi-condensed formula)

## Question

13 Draw the structural formulae for the following compounds: **a)** butanal; **b)** ethoxyethane; **c)** 2-chloropropanoic acid; **d)** propyl methanoate; **e)** pentan-2-one; **f)** 3,3-dimethylbutan-2-ol.

## 7.3 Properties of organic compounds

**DP link**

10 Organic chemistry

Physical and chemical properties of organic compounds depend on the nature of functional groups present in their molecules. Hydrocarbons and other non-polar substances tend to have lower melting and boiling points than compounds with polar functional groups. Unsaturated molecules, such as alkenes and alkynes, readily undergo addition reactions while substitution reactions are typical for molecules with no multiple carbon–carbon bonds. At the same time, combustion reactions are known for nearly all classes of organic compounds. In this topic, we will discuss a few reactions of each kind.

## Alkanes and halogenoalkanes

Alkanes are saturated hydrocarbons that contain only single C–H and C–C covalent bonds. Since carbon and hydrogen have similar electronegativities (2.6 and 2.2, respectively), these bonds have very low polarity, so the molecules of alkanes are held together only by weak London dispersion forces. As a result, short-chain alkanes under normal conditions are either gases (methane to butane) or volatile liquids (pentane and hexane). Melting and boiling points of alkanes increase with chain length (figure 5).

 **Internal link**

Electronegativity and London dispersion forces are described in **2.3 Chemical bonding**.

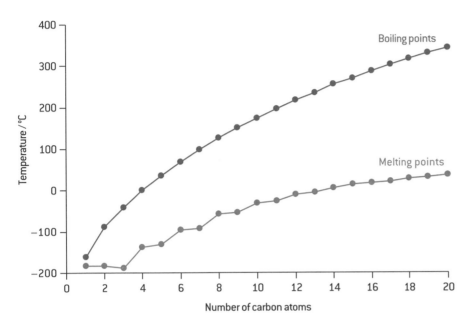

**Figure 5.** Melting and boiling points of alkanes

Because of their low polarity, alkanes are almost insoluble in water. At the same time, liquid alkanes readily mix with other non polar substances and are often used as organic solvents.

Carbon–carbon and carbon–hydrogen bonds in alkanes are relatively strong (table 4 in *5.1 Thermochemistry*), so these hydrocarbons have low chemical activity. Under normal conditions, alkanes are resistant to acids, bases and all common oxidizing and reducing agents. At higher temperatures, alkanes become more reactive and can participate in highly exothermic combustion reactions. In excess oxygen, alkanes undergo *complete combustion*, producing carbon dioxide and water:

$$CH_4(g) + 2O_2(g) \rightarrow CO_2(g) + 2H_2O(l) \qquad \Delta H^\circ = -891.1 \text{ kJ}$$
$$C_6H_{14}(l) + 9.5O_2(g) \rightarrow 6CO_2(g) + 7H_2O(l) \qquad \Delta H^\circ = -4163 \text{ kJ}$$

In actual combustion reactions, water vapour (steam) is often formed instead of liquid water, which makes these reactions less exothermic.

*Incomplete combustion* occurs when the supply of oxygen is limited. Depending on the alkane to oxygen ratio, incomplete combustion can produce carbon monoxide or elementary carbon (soot):

$$CH_4(g) + 1.5O_2(g) \rightarrow CO(g) + 2H_2O(l) \qquad \Delta H^\circ = -608.1 \text{ kJ}$$
$$CH_4(g) + O_2(g) \rightarrow C(s) + 2H_2O(l) \qquad \Delta H^\circ = -497.6 \text{ kJ}$$

 **Internal link**

Bond enthalpies and enthalpies of reaction are covered in **5.1 Thermochemistry**.

 **Key term**

**Complete combustion** of a hydrocarbon produces carbon dioxide and water.

**Incomplete combustion** of a hydrocarbon produces water and carbon monoxide or elementary carbon.

Carbon monoxide is highly toxic, as it binds to hemoglobin in the blood and thus prevents it from carrying oxygen to body tissues. Accidental exposure to carbon monoxide produced by incomplete combustion of fuels is a major cause of fatal poisoning in many countries.

### Question

14 State and balance the equations for complete and incomplete combustion of propane.

---

**DP ready** | **Nature of science**

### Fossil fuels

Alkanes are major components of the crude oil and natural gas that remain the principal sources of energy for industry. The burning of fossil fuels releases large amounts of carbon dioxide into the atmosphere and leads to global warming. In addition, fossil fuels are non-renewable, so their depletion increases the cost of energy and limits the supply of raw materials for chemical industry. Therefore, it is critically important to reduce the consumption of fossil fuels and use alternative sources of energy, such as biofuels, solar, wind and nuclear power.

---

**DP link**

You will learn more about these and other technologies by studying C Energy.

---

Reactions of alkanes with molecular chlorine and bromine produce halogenoalkanes and hydrogen halides; for example:

$$CH_4(g) + Cl_2(g) \xrightarrow{h\nu} CH_3Cl(g) + HCl(g)$$

The symbol "$h\nu$" over the reaction arrow refers to ultraviolet (UV) light, which is required to break the halogen molecule into free radicals and initiate the reaction:

$$Cl_2(g) \xrightarrow{h\nu} 2Cl\cdot(g)$$

Alternatively, halogen radicals can be produced by heating the reaction mixture. Once formed, these unstable and highly reactive species attack the alkane molecule, removing a hydrogen atom and producing an alkyl radical:

$$CH_4(g) + Cl\cdot(g) \rightarrow H_3C\cdot(g) + HCl(g)$$

In turn, the alkyl radical attacks a halogen molecule, producing halogenoalkane and regenerating the halogen radical:

$$H_3C\cdot(g) + Cl_2(g) \rightarrow CH_3Cl(g) + Cl\cdot(g)$$

The last two steps are repeated over and over again, so a single radical formed in the reaction mixture can produce many thousands of molecules of halogenoalkane and hydrogen halide. This process, known as a *chain reaction*, is terminated when two radicals recombine with one another and form a stable molecule:

$$H_3C\cdot(g) + Cl\cdot(g) \rightarrow CH_3Cl(g)$$

$$2Cl\cdot(g) \rightarrow Cl_2(g)$$

$$2H_3C\cdot(g) \rightarrow C_2H_6(g)$$

---

**Key term**

**Chain reactions** are cyclic processes that involve free radicals or other unstable species.

Note that the combination of two alkyl radicals produces a new alkane (in our case, ethane) with twice as many carbon atoms as that in the original molecule.

**Question**

15 State the equation for the reaction of ethane with molecular bromine. Outline the reaction mechanism and explain why small quantities of butane are formed as a by-product in this reaction.

In contrast to chlorine and bromine, the reactivity of iodine is insufficient for substituting hydrogen atoms in alkanes. On the other hand, fluorine is too active, so its reactions with alkanes often lead to explosions and form complex mixtures of products.

Halogenoalkanes are more reactive than alkanes due to the presence of polar carbon–halogen bonds. These bonds are easily broken in reactions with electron-rich species, such as hydroxide ions or ammonia. For example, a dilute aqueous solution of sodium hydroxide is commonly used for converting halogenoalkanes into alcohols:

$$CH_3Cl(g) + NaOH(aq) \rightarrow CH_3OH(aq) + NaCl(aq)$$

This equation can also be written in ionic form:

$$CH_3Cl(g) + OH^-(aq) \rightarrow CH_3OH(aq) + Cl^-(aq)$$

Similarly, reactions of halogenoalkanes with ammonia produce amines:

$$CH_3Cl(g) + 2NH_3(aq) \rightarrow CH_3NH_2(aq) + NH_4Cl(aq)$$

**Question**

16 State the equations for the reactions of 2-bromopropane with:
a) dilute aqueous solution of sodium hydroxide;
b) excess aqueous solution of ammonia.

### Alkenes and alkynes

Alkenes and alkynes are unsaturated hydrocarbons with double and triple carbon–carbon bonds, respectively. The second and third covalent bonds in multiple carbon–carbon bonds are slightly weaker than a single carbon–carbon bond, so these "extra" bonds break easily and provide free electrons that can be shared with other atoms. As a result, unsaturated hydrocarbons undergo various addition reactions. Some of these are listed below.

- *Hydrogenation* (reaction with hydrogen): these reactions are catalysed by transition metals, such as nickel or platinum, and usually require high temperature and pressure. For example, the hydrogenation of ethene proceeds as follows:

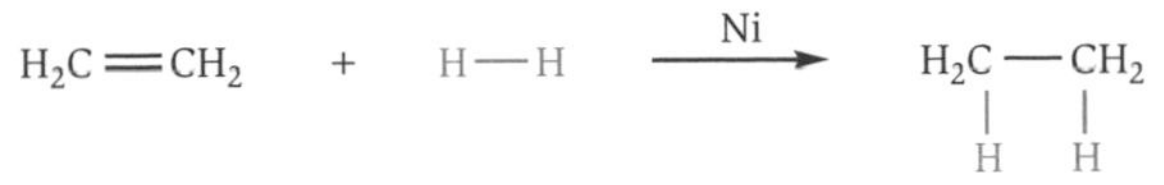

Under the same reaction conditions, alkynes undergo a stepwise hydrogenation, producing first alkenes and then alkanes:

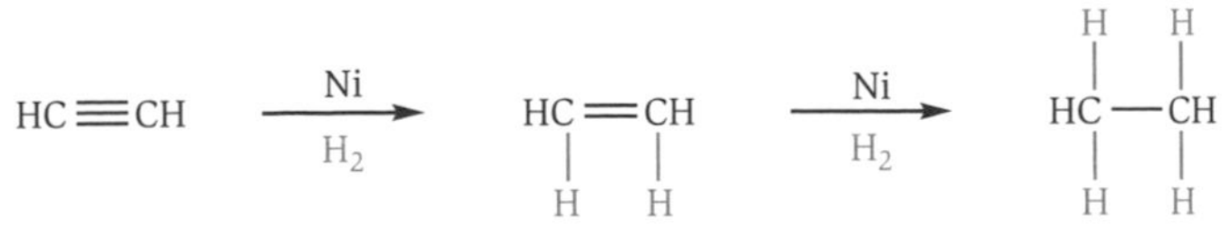

If the alkyne is taken in excess, the hydrogenation can be stopped at the alkene stage. When excess hydrogen is used, the alkane will be the only reaction product.

- *Halogenation* (reaction with halogens): these reactions do not require a catalyst and proceed readily even at room temperature, producing dihalides (or tetrahalides from alkynes):

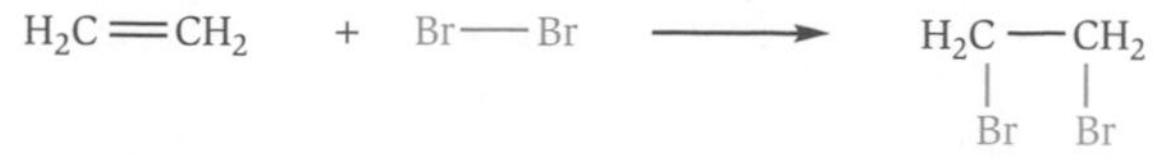

$$H_2C = CH_2 \ + \ Br—Br \ \longrightarrow \ \underset{\overset{|}{Br} \ \ \overset{|}{Br}}{H_2C—CH_2}$$

- *Hydrohalogenation* (reaction with hydrogen halides): these reactions proceed without a catalyst and produce halogenoalkanes.

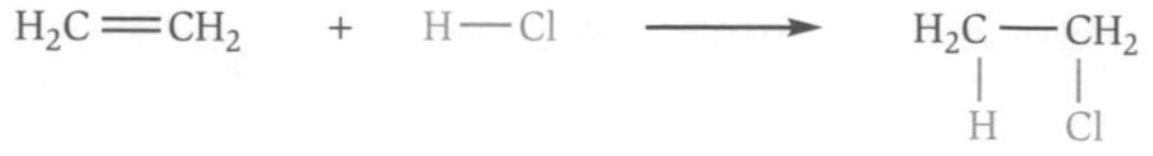

$$H_2C = CH_2 \ + \ H—Cl \ \longrightarrow \ \underset{\overset{|}{H} \ \ \overset{|}{Cl}}{H_2C—CH_2}$$

- *Hydration* (reaction with water): these reactions are catalysed by inorganic acids, such as $H_3PO_4$ or $H_2SO_4$, and produce alcohols:

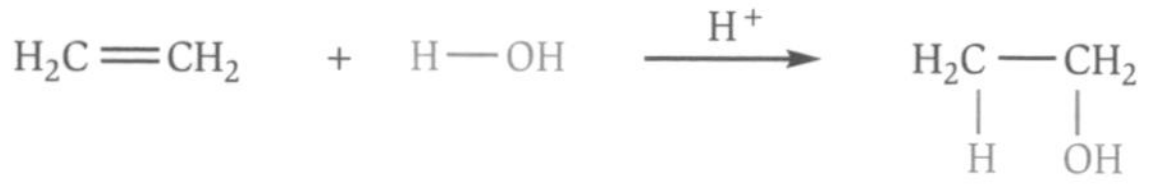

$$H_2C = CH_2 \ + \ H—OH \ \xrightarrow{H^+} \ \underset{\overset{|}{H} \ \ \overset{|}{OH}}{H_2C—CH_2}$$

Although aromatic hydrocarbons are unsaturated, they do not readily undergo addition reactions. The aromatic ring is very stable, so the hydrogenation of benzene and its homologues proceeds more slowly and releases less heat than the hydrogenation of alkenes and alkynes. Instead, typical reactions of the benzene ring involve the substitution of hydrogen atoms on the ring by functional groups or halogens.

## Question

17 State the equations for the following reactions: **a)** propene with hydrogen; **b)** but-1-ene with bromine; **c)** excess propyne with hydrogen; **d)** but-2-yne with excess chlorine. State the reaction conditions (where necessary) and deduce the IUPAC names of organic products.

## Practical skills: The bromine test

An aqueous solution of bromine, commonly called "bromine water", can be used for distinguishing saturated and unsaturated organic compounds. In a typical experiment, a small quantity of the analysed compound is added to bromine water, and the mixture is shaken for several seconds. If the characteristic orange-brown colour of bromine water disappears, the analysed compound is likely to contain a carbon–carbon double or triple bond. If the colour remains unchanged, the compound is saturated.

## Question

18 State the equations for the reactions of but-2-ene with hydrogen bromide and water. Where necessary, state the reaction conditions. In all cases, deduce the IUPAC names of organic products.

In the absence of other reactants, unsaturated organic molecules can join together and form *polymers*. The most common polymer, polyethene, is produced by addition polymerisation of ethene:

$$n \, H_2C{=}CH_2 \rightarrow \{H_2C{-}CH_2\}_n$$

ethene          polyethene

The average number of repeating units per macromolecule ($n$) is called the *degree of polymerisation*. For a typical polymer, the value of $n$ can vary from several hundred to tens of thousands.

Polymerisation of alkenes proceeds at high temperatures and requires a source of free radicals or other catalyst to break a carbon–carbon bond in the first molecule. Once the first molecule has become a free radical, the process continues as a chain reaction much like the halogenation of alkanes.

Other unsaturated compounds polymerize in the same way, producing macromolecules with various side chains:

chloroethene
(vinyl chloride)

polychloroethene
(polyvinyl chloride, PVC)

2-methylpropene
(isobutylene)

poly(2-methylpropene)
(polyisobutylene)

In all cases, the structure of the repeating unit reflects the structure of the monomer, with the double carbon–carbon bonds replaced by a single bond.

### Question

19 Draw the structures of the following polymers:
   a) polypropene (PP)
   b) polytetrafluoroethene (PTFE, also known as Teflon®)
   c) poly(but-2-ene).

## Alcohols and carbonyl compounds

Alcohols are organic compounds with a hydroxyl (–OH) functional group. In contrast to hydrocarbons, the molecules of alcohols are polar and can form hydrogen bonds with one another and with water. As a result, alcohols are soluble in water and have relatively high melting and boiling points (figure 6). Under normal conditions, short-chain alcohols are clear colourless liquids with characteristic "biting" odours.

---

**Key term**

A **polymer** is a large molecule (**macromolecule**) which is composed of many repeating units (**monomers**) connected to one another by covalent bonds.

**DP link**

A.5 Polymers

| DP ready | Nature of science |

**Polymers and the environment**

Polymers form the basis of the plastics industry, which provides a broad range of products for all kinds of applications — from packaging and toys to cars and satellites. The ever-increasing use of plastics improves the quality of our lives but at the same time causes serious environmental problems, such as general pollution and the extinction of marine animals.

**Internal link**

Hydrogen bonding is discussed in 2.3 Chemical bonding.

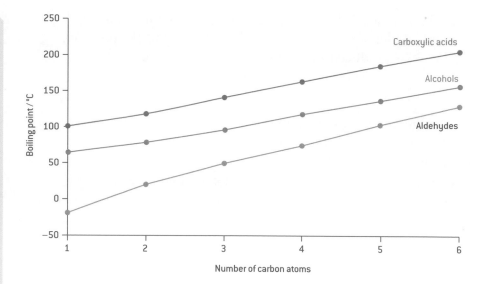

**Figure 6.** Boiling points of alcohols, aldehydes and carboxylic acids

The presence of a hydroxyl group enables alcohols to participate in various chemical reactions, including oxidation and condensation. The most common reagent used for oxidation of alcohols is acidified aqueous solution of potassium dichromate ($K_2Cr_2O_7$). A boiling mixture of this solution with an alcohol gradually changes colour from orange to green as dichromate ions are reduced to chromium(III) ions:

$$Cr_2O_7{}^{2-}(aq) + 14H^+(aq) + 6e^- \rightarrow 2Cr^{3+}(aq) + 7H_2O(l)$$

orange    green

In reaction schemes, the oxidizing agent is often denoted as "[O]" above the reaction arrow. This allows chemists to emphasize the changes in the molecules of organic compounds, which are usually more important than inorganic reagents.

Ethanol ($CH_3CH_2OH$) and other alcohols with a $CH_2OH$ fragment (called *primary* alcohols) are oxidized by acidified potassium dichromate stepwise, producing first aldehydes and then carboxylic acids:

ethanol      ethanal      ethanoic acid

Aldehydes have lower boiling points than both alcohols and carboxylic acids (figure 6), as their molecules cannot form hydrogen bonds with one another. Therefore, an aldehyde can be quickly removed from the reaction mixture by *distillation*. In contrast, if we need to synthesize a carboxylic acid, we must keep the aldehyde in the reaction mixture for a longer period of time. This can be achieved by heating the mixture under *reflux*.

Alcohols with a CHOH fragment (called *secondary* alcohols) are oxidized to ketones:

propan-2-ol      propanone

The carbon atom of the C=O group in ketones has no hydrogen atoms that could be removed or replaced with oxygen. Because of that, ketones cannot be oxidized any further without breaking carbon–carbon bonds, which does not occur under typical reaction conditions. For the same reason, alcohols with no C–H bonds adjacent to the hydroxyl group are also resistant to oxidation and do not react with acidified potassium dichromate.

## Question

**20** Deduce the formulae and names of organic products formed in the reactions of acidified potassium dichromate with the following compounds: **a)** butan-1-ol; **b)** butan-2-ol; **c)** 2-methylbutan-2-ol; **d)** butanal.

Aldehydes, ketones and carboxylic acids are collectively known as *carbonyl compounds,* as their molecules contain a carbonyl (C=O) group. Carboxylic acids and alcohols undergo *condensation reactions* to give another class of carbonyl compounds, esters; for example:

ethanoic acid    methanol    methyl ethanoate

Reactions of carboxylic acids with alcohols are catalysed by concentrated inorganic acids, such as $H_2SO_4$, HCl or $H_3PO_4$. Since esterification is a reversible process, only a certain proportion of the reactants can be converted into the products before the equilibrium is reached. According to Le Châtelier's principle, the reaction yield can be increased by using an excess of one of the reactants (typically, the cheapest) or removing one of the products from the reaction mixture. Many esters have relatively low boiling points, as their molecules cannot form hydrogen bonds with one another. Such esters can be distilled off from the reaction mixture, shifting the equilibrium position to the right. For longer-chain esters with high boiling points, the same effect can be achieved by distilling off water.

## Question

**21** State the equation for the esterification of methanoic acid with propan-2-ol. State the reaction conditions and the IUPAC name of the organic product.

Condensation reactions are very common in organic and biological chemistry. In addition to alcohols and carboxylic acids, they can involve other classes of organic compounds, such as amines and amino acids.

### Key term

**Carbonyl compounds** contain a C=O group in their molecules.

In a **condensation** reaction, two or more organic molecules join together ("condense") into a larger molecule (main product) and release a smaller, usually inorganic molecule (by-product). A reaction between a carboxylic acid and an alcohol is called **esterification**, as it produces an ester as the main condensation product.

### Internal link

Le Châtelier's principle is described in **5.3 Chemical equilibrium**

### DP link

Condensation reactions of amino acids are discussed in **B.2 Proteins and enzymes**.

## Chapter summary

In the final chapter of this book, you have learned about the classification, nomenclature and properties of organic compounds. By now, you should have a working knowledge of the following concepts and definitions:

☐ The carbon atom is tetravalent in almost all organic compounds.

☐ Structures of organic molecules can be represented by full, condensed, semi-condensed and skeletal formulae.

☐ Shapes of molecules can be visualized by 3D models.

☐ The VSEPR theory can be used to predict the bond angles in simple organic molecules.

☐ Organic compounds are divided into classes according to functional groups in their molecules.

☐ A homologous series contains compounds with similar structures that differ by one or more $CH_2$ groups.

☐ Structural isomers have the same molecular formula but different structural formulae.

☐ Each organic compound has a specific IUPAC name, which is constructed systematically.

☐ All members of a homologous series have similar chemical properties but differ in physical properties, which change gradually as the chain length increases.

☐ In a series of compounds with similar molecular masses, the melting and boiling points generally increase in the following order: hydrocarbon < aldehyde ≈ ketone < alcohol < carboxylic acid.

☐ Complete combustion of organic compounds produces carbon dioxide, water and, in certain cases, other products, such as molecular nitrogen.

☐ Incomplete combustion produces carbon monoxide and/or elementary carbon (soot) instead of carbon dioxide.

☐ Alkanes undergo substitution reactions with halogens, catalysed by UV light and involving free radicals as intermediates.

☐ Halogenoalkanes react with a dilute solution of sodium hydroxide to produce alcohols and with ammonia to produce amines.

☐ Alkenes undergo addition reactions with hydrogen, hydrogen halides, halogens and water to produce alkanes, halogenoalkanes, dihalogenoalkanes and alcohols, respectively.

☐ Alkynes react with hydrogen and halogens stepwise, producing first unsaturated and then saturated compounds.

☐ Saturated and unsaturated hydrocarbons can be distinguished by the bromine test.

☐ Alkenes undergo addition polymerisation in the presence of free radicals or other catalysts.

☐ Benzene undergoes substitution reactions more readily than addition reactions.

☐ Primary alcohols are oxidized by acidified potassium dichromate stepwise, producing first aldehydes and then carboxylic acids.

☐ The degree of oxidation of alcohols can be controlled by using distillation or reflux.

☐ Secondary alcohols are oxidized to ketones while tertiary alcohols are resistant to oxidation.

☐ Esters can be synthesized by condensation reactions between carboxylic acids and alcohols.

## Additional problems

1. Copy and complete the table below.

| Name | Structural formula | | |
| | Full | Condensed | Skeletal |
| propyne | | | |
| | H—C—C⟍ with H, H, =O, OH (full structure shown) | | |
| | | $(CH_3)_3COH$ | |
| | | | (skeletal structure with Cl) |

2. Deduce the structural formulae and systematic names for all structural isomers of alkenes with the molecular formula $C_5H_{10}$.

3. Draw the structural formulae for the following compounds: a) 3-methylpent-2-ene; b) 1,1,3-trichloropropane; c) 4-aminobutanoic acid; d) 2-ethoxybutane.

4. Deduce the general formulae for the following homologous series of organic compounds with no multiple carbon–carbon bonds: a) alcohols; b) aldehydes; c) carboxylic acids. For each series, state one other class of organic compounds with the same general formula.

5. State and balance the equations for complete combustion of the following organic compounds: a) butane; b) methyl ethanoate; c) trimethylamine.

6. Benzene is a component of petrol (gasoline) and other common fuels. a) State and balance the equations for complete and incomplete combustion of benzene. b) Using table 1 from *5.1 Thermochemistry* and the $\Delta H_f^\theta$ value for liquid benzene ($+49.0$ kJ mol$^{-1}$), calculate the standard enthalpy changes for these combustion reactions.

7. Bromomethane can be synthesized from methane and molecular bromine. State the conditions required for this reaction and outline its mechanism.

8. Two unlabelled bottles contain the following solutions: a) pent-1-ene in pentane; b) benzene in hexane. Suggest how these solutions can be distinguished using a single reagent (an individual substance or mixture). State the equations for possible chemical reactions and describe the changes observed in each case.

9. Hot drinks are often sold in styrofoam cups, which are made of polystyrene. This material is produced by the addition polymerisation of styrene, $C_6H_5CH=CH_2$. a) Deduce the structure of polystyrene. b) Calculate the average molecular mass of polystyrene if its degree of polymerisation is 4500.

10. State the equations for the reactions of chloroethane with: a) dilute aqueous solution of sodium hydroxide; b) excess aqueous solution of ammonia.

11. Explain why aldehydes and ketones have higher boiling points than hydrocarbons but lower boiling points than alcohols and carboxylic acids with similar molecular masses.

12. Deduce which of the following pairs of organic compounds can be distinguished using an acidified solution of potassium dichromate: a) pentan-1-ol and pentan-2-ol; b) pentan-2-ol and 2-methylpentan-2-ol; c) pentanal and pentan-2-one; d) 3-methylpentan-3-ol and pentan-3-one.

13. State the equation for the esterification of pentanoic acid with pentan-1-ol. State the reaction conditions and the IUPAC name of the organic product. Suggest how the yield of this product can be increased if the boiling points of pentanoic acid, pentan-1-ol and the organic product are 185, 138 and 207°C, respectively.

# 8 Tips and advice on successful learning

## 8.1 Approaches to your learning

Every science course, including DP Chemistry, requires a wide range of study skills. Many of these skills are known as **Approaches to learning**. As discussed in the *Introduction*, you will be encouraged to develop these skills throughout the diploma programme.

- **Communication skills** are necessary for effective exchange of knowledge through interaction. These skills allow you to give and receive meaningful feedback, collaborate with peers and experts, interpret communication through intercultural understanding and use appropriate forms of presenting information for different audiences.

  Effective communication is particularly important in science, where the ideas and theories are constantly discussed and scrutinized. Lack of communication slows down the progress of science by forcing researchers to repeat the same work over and over again. For example, the discovery of the law of conservation of mass by Mikhail Lomonosov in 1748 (*1.3 Stoichiometric relationships*) was published only in Russian journals and remained unknown to European scientists for over 25 years, until Antoine Lavoisier formulated the same principle in 1774.

- **Social skills** are closely related to communication and involve establishing and maintaining relationships, respecting others, taking and sharing responsibility, managing and resolving conflicts, reaching consensus and making fair Decisions. The development of these skills is an important goal of the interdisciplinary project that requires close collaboration between students of different subjects.

  Many social skills go beyond interpersonal relationships and define our ability to tackle social and environmental problems at local, national and international levels. The concept of "green chemistry" (*1.3 Stoichiometric relationships*) is an example of a responsible and ethical approach to research and technology.

- **Self-management** skills facilitate study and research by careful planning and organization, mental concentration, persistence, emotional control, self-motivation, resilience and reflection. This set of skills is crucial for successful learning, as the intensive nature and diversity of DP courses require effective management of time and tasks, the ability to meet the deadlines, overcome challenges and reflect on inevitable failures and errors.

  The persistence, resilience and reflection skills of individual researchers are closely related to the "risk-taking" nature of science, where the hypotheses and theories are often disproved or superseded. Dalton's atomic theory (*1.1 The particulate nature of matter*) and Mendeleev's periodic law (*2.2 The periodic law*) were imperfect and had to be amended before they were accepted by the scientific community.

- **Research skills** include finding, interpreting and analysing information, collecting and processing experimental data, developing long-term memory and using a variety of multimedia resources and digital tools for problem-solving and presentation purposes. These skills are at the heart of inquiry-based IB programmes, including the DP Chemistry course. Personal research is an integral part of the extended essay, which provides the opportunity for you to develop research skills with the support and guidance of a supervisor.

  The importance of research skills can be illustrated by almost every breakthrough in science. For example, the planetary model of the atom (*1.1 The particulate nature of matter*) was based on careful analysis of all scientific information available at that time and required new experimental data (Rutherford's gold foil experiment) to be collected, processed and interpreted using advanced mathematical techniques.

  Certain research skills are closely connected to academic honesty. All DP students are required to acknowledge the work of other people incorporated in their essays or presentations. The lack of citation and referencing skills may lead to unintentional malpractice and result in a penalty imposed by the IB award committee.

- **Thinking skills** are a diverse set of skills that can be subdivided into three large categories: critical thinking, creativity and innovation, and transfer.

  - *Critical thinking* involves analysing and evaluating issues and ideas by observation, interpretation and generalization. In addition to gathering and organizing information, both students and researchers must be able to formulate and evaluate arguments, recognize unstated assumptions and bias, draw reasonable conclusions, consider ideas from multiple perspectives and use models or simulations to explore complex systems. The discovery of the periodic law by Mendeleev and its further development by Moseley (*2.2 The periodic law*) are good examples of critical thinking, as both cases involved generalization and conclusions based on observation and evaluation of experimental data.

  - *Creativity and innovation* are the skills of invention that allow us to develop ideas and material objects that never existed before. These skills often require considering unlikely or seemingly impossible solutions, making guesses and generating testable hypotheses, using lateral thinking for making unexpected connections and proposing improvements to existing machines and technologies. Although creativity and innovation are usually associated with scientific discoveries and inventions, you also need to develop these skills in order to produce original work and prepare yourself for a career in science, industry or education.

  - *Transfer skills* are particularly important in education, as they allow you to utilize effective learning strategies in subject groups and disciplines, apply existing skills and knowledge in unfamiliar situations, change the context of an inquiry to gain different perspectives, and combine knowledge, understanding and skills to create new solutions and products. The ability to "think in context" opens new dimensions in our perception of science and technology, including their social, political, economic and environmental implications. For example, the Haber process (*5.3 Chemical equilibrium*) can be viewed as an illustration of Le Châtelier's principle but at the same time as the primary source of nitrogen for agriculture and the most important discovery in the 20th century.

The five groups of approaches to learning skills, together with six groups of approaches to teaching (inquiry-based, conceptually focused, contextualized, collaborative, differentiated and informed by assessment) encompass the key values and principles of IB teaching. Students and teachers are expected to work together to identify and build up additional subject-specific skills that can facilitate the learning process and ultimately achieve the primary goal of the IB mission statement of developing "inquiring, knowledgeable and caring young people who help to create a better and more peaceful world through intercultural understanding and respect."

## 8.2 Good study habits

In addition to the **Approaches to learning** skills, here are some simple recommendations you can apply to your learning process to make it less challenging and more satisfying.

1. **Get ready for study.** Have enough sleep, eat well, drink plenty of water and reduce your stress with positive thinking and physical exercise.

2. **Organize your study environment.** Find a comfortable place with adequate lighting, temperature and ventilation. Eliminate all possible distractions by switching off your phone, TV and radio. Keep your papers and computer files organized. Bookmark usvful online resources and back up your data regularly.

3. **Plan your studies.** Make a list of your tasks and arrange them by importance. Break up large tasks into smaller, easily manageable parts. Create an agenda for your studying time and make sure that you can complete each task before the deadline.

4. **Read actively.** Identify the purpose of reading and get an idea of what you are going to learn. Skim through the text and figures, concentrating on titles, headings and boldface words. Write down any questions that might come to your mind. Read the text and try to answer these questions. Take notes for future reference.

5. **Build up your knowledge.** Focus on understanding rather than memorizing. Recite key points and definitions using your own words. Organize your memory by making links between the facts and concepts you need to remember. Use visualization, associations and mnemonics whenever possible. Try to apply theoretical concepts to practical problems.

6. **Develop problem-solving skills.** Answer in-text questions and do self-study exercises. Do not look at the answer key until you have solved the problem yourself. Do not give up until you have exhausted all options, including online search for the solutions to similar problems.

7. **Practice critical thinking.** Question all ideas, concepts and assumptions instead of accepting them at face value. Identify inconsistencies and errors in reasoning. Do not accept vague explanations. Approach problems systematically and from different perspectives. Engage in discussions. Justify your own values and beliefs.

8. **Be curious.** Do not limit your studies to specific assignments; read around. Make a habit of learning something new every day. Do not hesitate to ask questions of your teacher and other students. Identify the areas of your personal interest within the course and study this material in depth. Engage in extracurricular activities and share your interests with other people.

9. **Prepare for the exams.** Read the syllabus and identify the gaps in your understanding and skills. Revise the topics where improvement is required. Practice answering exam-style questions under a time constraint. Solve all problems from past papers and take a mock exam if possible.

10. **Do not panic.** Take a positive attitude and concentrate on things you can improve. Set realistic goals and work systematically to achieve these goals. Be prepared to reflect on your performance and learn from your errors in order to improve your future results.

## 8.3 Academic honesty

Throughout your Diploma Programme, not just in chemistry, you will be collecting the words and thoughts of others, from the opinions of classmates through to the writings of distinguished scientists. It is entirely appropriate that you should quote other people in support of your work. What is not appropriate is passing these words off as your own to gain credit. This is academic dishonesty and, if you are quoting verbatim from others without crediting them, is plagiarism. The quote at the beginning of Chapter 3 of this book has a scientific bibliographic reference.

## 8.4 Glossary of command terms

The key to answering an examination question well is to read it carefully. Pay special attention to the verb, known as the command term, that indicates how you should answer. This command term is normally at the start of the question sentence. Here are some of the more common terms used and their meaning:

| | |
|---|---|
| **Calculate** | Obtain a numerical answer showing your working. |
| **Classify** | Arrange or order by class or category. |
| **Compare** | Give an account of the similarities and differences between two or more items or situations. |
| **Deduce** | Reach a conclusion from the information given. |
| **Define** | Give the precise meaning of a word, phrase, concept or physical quantity. |
| **Describe** | Give a detailed account. |
| **Determine** | Obtain the only possible answer. |
| **Discuss** | Offer a review that includes a range of arguments, factors or hypotheses. Opinions or conclusions should be presented clearly and supported by appropriate evidence. |
| **Draw** | Represent by means of a labelled diagram or graph, using a pencil. |
| **Estimate** | Obtain an approximate value. |
| **Examine** | Consider an argument or concept in a way that uncovers the assumptions and interrelationships of the issue. |

| | |
|---|---|
| **Explain** | Give a detailed account including reasons or causes. |
| **Explore** | Undertake a systematic process of discovery. |
| **Identify** | Provide an answer from a number of possibilities. |
| **Label** | Add labels to a diagram. |
| **Outline** | Give a brief account or summary. |
| **Predict** | Give an expected result. |
| **Sketch** | Represent by means of a labelled diagram or graph. The sketch should give a general idea of the required shape or relationship. |
| **State** | Give a specific name, value or other brief answer without explanation or calculation. |
| **Suggest** | Propose a solution, hypothesis or other possible answer. |

## 8.5 Internal assessment

Details of the *internal assessment* (IA) were given in the introduction. You carry this 10-hour project through by yourself or in a small group as an integral part of the course. You will carry out the data analysis by yourself and must not collude with other students, even though collaboration is allowed in the early stages of the IA.

The type of project that can be undertaken is very broad and might include:

- traditional experimental data collection
- investigation of a database
- processing a spreadsheet
- investigating the behaviour of a simulation.

IAs are marked by your teacher and then sent electronically to the IB for moderation. Moderation means that some or all the work of your student group will be marked again by an independent assessor who checks that your school's marking standards match those of other schools.

Each IA project is scored out of 24 marks. The criteria used for marking are

| Criterion | Description | Weighting |
|---|---|---|
| *Personal engagement* | This measures how well you engage in the project. It may be a personal interest, or you may be especially attracted to the topic for some reason. Your thinking about the topic may have been particularly creative, or you may have personalized how you expressed scientific concepts. Try to make the project your own. | 8% |
| *Exploration* | To score highly here you need a research question that is focused, and you must identify the issues that surround it. The concepts and techniques should arise naturally from the topic. | 25% |
| *Analysis* | You should process, analyse and interpret your data to arrive at a detailed and valid answer to your research question. Your data analysis must be accurate and there must be sufficient data to justify the conclusion. It will be important to consider experimental uncertainty in your conclusion. | 25% |
| *Evaluation* | You must evaluate the quality of the experimental and analytical work you did. It may be appropriate to compare your results with the published work of scientists or other accepted results. You should suggest improvements to your research and should draw attention to any shortcomings in it. | 25% |
| *Communication* | Your report should be concise and effective. The presentation of text, tables, diagrams and graphs should be of a high quality. The overall length of the main text should not be greater than about 12 pages. | 17% |

As you progress through your IA, keep referring to the marking criteria to ensure that you are producing what is required from this coursework.

## 8.6 Working with mathematical expressions

### Rearranging formulae

Many mathematical expressions in this book are given in the form $a = \dfrac{b}{c}$. To find $b$ or $c$, you need to rearrange the equation as follows:

$$b = a \times c \qquad c = \dfrac{b}{a}$$

Alternatively, you can use the formula triangle (figure 1). To find a variable, close it in the triangle with your finger. The remaining two variables will form the correct mathematical expression:

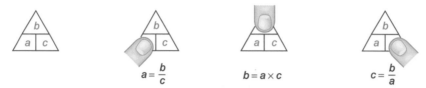

**Figure 1.** How to use a formula triangle

To make the triangle from an expression like $b = a \times c$, take the two quantities that are multiplied together and put them in the bottom sections of the triangle; then put the term on the other side of the equation on the top. If the expression is in the form $a = \dfrac{b}{c}$, put the term from the top of the fraction in the top of the triangle.

### Squares and square roots

Squares and square roots are inverse mathematical operations. For example, if $a = b^2$, then $b = \sqrt{a}$. You will need a scientific calculator to find square roots of physical quantities.

A square can be calculated by multiplying the quantity by itself: $b^2 = b \times b$.

For example, $10^2 = 10 \times 10 = 100$.

### Logarithms

Logarithms are often used for expressing very large or very small numbers, such as concentrations of hydrogen ions in aqueous solutions (see *6.3 The concept of pH*). The common, or decimal, logarithm of a given number $a$ is the power to which the logarithm base (usually 10) must be raised to produce this number.

Therefore, if $\log_{10}(a) = b$, then $b = 10^a$.

In mathematical expressions, the base 10 is often omitted, so "$\log_{10}(a)$" is usually written as "$\log(a)$".

As an example, let's calculate the common logarithm of 100. This number is the square of the logarithm base (10):

$10^2 = 100$.

Therefore, $\log(100) = 2$.

Similarly, $\log(1000) = 3$, as $10^3 = 1000$, and $\log(263) = 2.42$, as $10^{2.42} = 263$.

In the general case, $\log(10^x) = x$.

To find a logarithm of an arbitrary number (not a power of ten), you will have to use a scientific calculator.

Negative logarithms are known as "p-numbers". For example, $pH = -\log[H^+]$, and $pK_w = -\log(K_w)$.

Logarithms and p-numbers are discussed in more detail in *6.3 The concept of pH* in this book.

# Appendix

**Table 1.** Base SI units

| Quantity | Symbol | Unit | Symbol for unit |
|---|---|---|---|
| length | $\ell$ | metre | m |
| mass | $m$ | kilogram | kg |
| time | $t$ | second | s |
| electric current | $I$ | ampere | A |
| temperature | $T$ | kelvin | K |
| amount of substance | $n$ | mole | mol |
| luminous intensity | $I_v$ | candela | cd |

**Table 2.** Other quantities and units

| Quantity | Symbol | Common units |
|---|---|---|
| relative atomic mass | $A_r$ | — |
| molar concentration (molarity) | $c$ | mol dm$^{-3}$ |
| enthalpy | $H$ | kJ *or* kJ mol$^{-1}$ |
| molar mass | $M$ | g mol$^{-1}$ |
| relative molecular mass | $M_r$ | — |
| number of species | $N$ | — |
| heat | $Q$ | kJ *or* kJ mol$^{-1}$ |
| entropy | $S$ | J K$^{-1}$ *or* J mol$^{-1}$ K$^{-1}$ |
| volume | $V$ | cm$^3$ *or* dm$^3$ |
| reaction rate | $v$ | mol dm$^{-3}$ s$^{-1}$ |
| density | $\rho$ | g cm$^{-3}$ *or* kg dm$^{-3}$ |
| mass percentage | $\omega$ | % |

**Table 3.** Decimal prefixes

| Text | Symbol | Factor |
|---|---|---|
| mega | M | $10^6$ |
| kilo | k | $10^3$ |
| deci | d | $10^{-1}$ |
| centi | c | $10^{-2}$ |

| Text | Symbol | Factor |
|---|---|---|
| milli | m | $10^{-3}$ |
| micro | μ | $10^{-6}$ |
| nano | n | $10^{-9}$ |
| pico | p | $10^{-12}$ |

**Table 4.** Physical constants

| Name | Symbol | Value and units |
|---|---|---|
| Avogadro's constant | $N_A$ | $6.02 \times 10^{23}$ mol$^{-1}$ |
| Gas constant | $R$ | $8.31$ J K$^{-1}$ mol$^{-1}$ |
| Molar volume of an ideal gas at STP (100 kPa and 273.15 K, or 0 °C) | $V_m$ | $22.7$ dm$^3$ mol$^{-1}$ |
| Molar volume of an ideal gas at SATP (100 kPa and 298.15 K, or 25 °C) | $V_m$ | $24.8$ dm$^3$ mol$^{-1}$ |
| Specific heat capacity of water | $C_w$ | $4.18$ kJ kg$^{-1}$ K$^{-1}$ |
| Ionic product of water at 25 °C | $K_w$ | $1.00 \times 10^{-14}$ mol$^2$ dm$^{-6}$ |
| Atomic mass unit | u, amu | $1.660540 \times 10^{-27}$ kg |
| Mass of proton | $m_p$ | $1.672622 \times 10^{-27}$ kg |
| Mass of neutron | $m_n$ | $1.674927 \times 10^{-27}$ kg |
| Mass of electron | $m_e$ | $9.109383 \times 10^{-31}$ kg |

**Table 5.** Useful equations

$$n = \frac{N}{N_A} \qquad n = \frac{m}{M} \qquad n = \frac{V}{V_m}$$

$$\rho = \frac{m}{V} \qquad M = \rho V_m \qquad pV = \frac{m}{M} RT$$

$$\omega = \frac{m_{solute}}{m_{solution}} \times 100\% \qquad c_{solute} = \frac{n_{solute}}{V_{solution}} \qquad Yield = \frac{n_{pract}}{n_{theor}} \times 100\%$$

$$Q = m_w C_w \Delta T \qquad \Delta H_c^o = -\frac{M}{m} \times Q$$

$$v_{avr} = \frac{|\Delta c|}{\Delta t} \qquad v_{inst} = \frac{|dc|}{dt}$$

$$pH = -\log[H^+] \qquad pOH = -\log[OH^-] \qquad pH + pOH = 14 \text{ (at 25 °C)}$$

$$\Delta H^{\ominus}(reaction) = \Sigma \Delta H_f^{\ominus}(products) - \Sigma \Delta H_f^{\ominus}(reactants)$$
$$\Delta H^{\ominus}(reaction) = \Sigma \Delta H_c^{\ominus}(reactants) - \Sigma \Delta H_c^{\ominus}(products)$$
$$\Delta H^{\ominus}(reaction) = \Sigma BE(bonds\ broken) - \Sigma BE(bonds\ formed)$$
$$\Delta H^{\ominus}(reaction) = \Sigma BE(reactants) - \Sigma BE(products)$$

**Table 6.** Relative electronegativities ($\chi$) of main-group elements

| | 1 | 2 | 3–12 | 13 | 14 | 15 | 16 | 17 | 18 |
|---|---|---|---|---|---|---|---|---|---|
| 1 | H 2.2 | | | | | | | | He — |
| 2 | Li 1.0 | Be 1.6 | | B 2.0 | C 2.6 | N 3.0 | O 3.4 | F 4.0 | Ne — |
| 3 | Na 0.9 | Mg 1.3 | | Al 1.6 | Si 1.9 | P 2.2 | S 2.6 | Cl 3.2 | Ar — |
| 4 | K 0.8 | Ca 1.0 | | Ga 1.8 | Ge 2.0 | As 2.2 | Se 2.6 | Br 3.0 | Kr — |
| 5 | Rb 0.8 | Sr 1.0 | | In 1.8 | Sn 2.0 | Sb 2.0 | Te 2.1 | I 2.7 | Xe — |
| 6 | Cs 0.8 | Ba 0.9 | | Tl 1.8 | Pb 1.8 | Bi 1.9 | Po 2.0 | At 2.2 | Rn — |

**Table 7.** Atomic radii ($r_a$) of main group elements / pm (1 pm = $10^{-12}$ m)

| | 1 | 2 | 3–12 | 13 | 14 | 15 | 16 | 17 | 18 |
|---|---|---|---|---|---|---|---|---|---|
| 1 | H 32 | | | | | | | | He 37 |
| 2 | Li 130 | Be 99 | | B 84 | C 75 | N 71 | O 64 | F 60 | Ne 62 |
| 3 | Na 160 | Mg 140 | | Al 124 | Si 114 | P 109 | S 104 | Cl 100 | Ar 101 |
| 4 | K 200 | Ca 174 | | Ga 123 | Ge 120 | As 120 | Se 118 | Br 117 | Kr 116 |
| 5 | Rb 215 | Sr 190 | | In 142 | Sn 140 | Sb 140 | Te 137 | I 136 | Xe 136 |
| 6 | Cs 238 | Ba 206 | | Tl 144 | Pb 145 | Bi 150 | Po 142 | At 148 | Rn 146 |

# Solubility of salts and hydroxides in water at 25°C

| | $NH_4^+$ | $Li^+$ | $Na^+$ | $K^+$ | $Ag^+$ | $Mg^{2+}$ | $Ca^{2+}$ | $Ba^{2+}$ | $Zn^{2+}$ | $Fe^{2+}$ | $Pb^{2+}$ | $Cu^{2+}$ | $Hg^{2+}$ | $Al^{3+}$ | $Cr^{3+}$ | $Fe^{3+}$ |
|---|---|---|---|---|---|---|---|---|---|---|---|---|---|---|---|---|
| $OH^-$ | — | s | s | s | — | i | ss | s | i | i | i | i | — | i | i | i |
| $F^-$ | s | ss | s | s | s | i | i | ss | ss | ss | i | s | ss | ss | i | ss |
| $Cl^-$ | s | s | s | s | i | s | s | s | s | s | ss | s | s | s | s | s |
| $Br^-$ | s | s | s | s | i | s | s | s | s | s | ss | s | ss | s | s | s |
| $I^-$ | s | s | s | s | i | s | s | s | s | s | i | — | i | s | s | — |
| $NO_3^-$ | s | s | s | s | s | s | s | s | s | s | s | s | s | s | s | s |
| $S^{2-}$ | s | s | s | s | i | — | s | s | i | i | i | i | i | — | — | — |
| $SO_3^{2-}$ | s | s | s | s | i | ss | i | i | i | i | i | i | i | — | — | — |
| $SO_4^{2-}$ | s | s | s | s | ss | s | ss | i | s | s | i | s | s | s | s | s |
| $PO_4^{3-}$ | s | i | s | s | i | i | i | i | i | i | i | i | i | i | i | i |
| $SiO_3^{2-}$ | — | i | s | s | i | i | i | i | i | i | i | i | — | i | i | i |
| $CO_3^{2-}$ | s | s | s | s | i | i | i | i | i | i | i | i | i | — | — | s |
| $CH_3COO^-$ | s | s | s | s | s | s | s | s | s | s | s | s | s | s | s | s |

*Notes:* 's' – soluble; 'ss' – slightly soluble; 'i' – insoluble; '—' – reacts with water or does not exist.

# Index

concentration expressions and stoichiometry 96–102

strong electrolytes 103–6

structural formulae 14

structural isomers 177

substitution reactions 79

surface area 130

surroundings 111

systematic names 15

systems 111

   closed systems 112

## T

temperature 131

   standard ambient temperature and pressure (SATP) 96

   standard temperature and pressure (STP) 22, 96

thermochemistry 111–12, 143–5

   bond enthalpy 121–2

   calculations involving standard enthalpy of combustion 116–18

   calculations involving standard enthalpy of formation 114–16

   calorimetry 118–21

   enthalpy changes 112–13

   entropy 122–3

   Hess's Law 113–14

thermodynamics, first law of 112

titration 102

toluene 181

total ionic equations 107

trivial names 15

## U

unsaturated hydrocarbons 180

unsaturated solutions 94

## V

valence 14

   valence electrons 36

   valence-shell electron-pair repulsion (VSEPR) theory 60–1

van der Waals forces 54

voltaic cells 86–7

## W

water 162–4

wave–particle duality 5

weak electrolytes 103–6

## X

xylene 181

## Y

yield 26–8